T0253930

Lecture Notes in Mathematics

An informal series of special lectures, seminars and reports on mathematical topics

Edited by A. Dold, Heidelberg and B. Eckmann, Zürich

15

F. Oort

Universiteit van Amsterdam

Commutative group schemes

1966

Springer-Verlag · Berlin · Heidelberg · New York

Library of Congress Catalog Card Number 66-20635. Title No. 7335.

Introduction

We restrict ourselves to two aspects of the field of group schemes, in which the results are fairly complete: commutative algebraic group schemes over an algebraically closed field (of characteristic different from zero), and a duality theory concerning abelian schemes over a locally noetherian prescheme. The preliminaries for these considerations are brought together in chapter I.

SERRE described properties of the category of commutative quasi-algebraic groups by introducing pro-algebraic groups. In characteristic zero the situation is clear. In characteristic different from zero information on finite group schemes is needed in order to handle group schemes; this information can be found in work of GABRIEL. In the second chapter these ideas of SERRE and GABRIEL are put together. Also extension groups of elementary group schemes are determined.

A suggestion in a paper by MANIN gave crystallization to a feeling of symmetry concerning subgroups of abelian varieties. In the third chapter we prove that the dual of an abelian scheme and the linear dual of a finite subgroup scheme are related in a very natural way. Afterwards we became aware that a special case of this theorem was already known by CARTIER and BARSOTTI. Applications of this duality theorem are: the classical duality theorem ("duality hypothesis", proved by CARTIER and by NISHI); calculation of $\mathrm{Ext}(\underline{\underline{G}}_a, A)$, where A is an abelian variety (result conjectured by SERRE); a proof of the symmetry condition (due to MANIN) concerning the isogeny type of a formal group attached to an abelian variety.

As we said before, our results originate from work of SERRE and GABRIEL. Besides that of course the ideas and results of GROTHENDIECK were indispensable. I am greatly indebted to J.-P.Serre, from whom I received valuable suggestions and helpful correspondence, to P.Gabriel, who, having read the manuscript, proposed many improvements and who gave precious information, and to J.P.Murre for his continuous interest in my work.

<u>Terminology and notations</u>. We say that a diagram is exact, if all columns and all rows are exact sequences. The terms injective, surjective and bijective are to be taken in the set-theoretical sense, while monomorphic, epimorphic and isomorphic are to be taken in the categorical sense (an epimorphic ring-homomorphism needs not to be surjective, a surjective morphism of schemes needs not to be an epimorphism). Working with preschemes, we use Mor(- , -) in order to indicate a set of morphisms (this notation differs from the corresponding one in EGA), while working with group schemes, we use Hom(- , -) in order to denote a set of homomorphisms (in this case HOM(- , -) seems to be the current notation). If E and F are modules over a ring A, we write $\mathrm{Hom}_A(E,F)$ for the set of A-homomorphisms from E into F. If B and C are A-algebras we write $\mathrm{RHom}_A(B,C)$ for the set of A-algebra homomorphisms from B into C. The multiplicative group of units of a ring A with identity element we denote by A^*. We use $X \in (Sch/S)$, or $X \in (Sch/A)$ in case $S = \mathrm{Spec}(A)$, as an abbreviation for "X is a prescheme over S"; we consider locally noetherian preschemes only, if not mentioned otherwise.

Square brackets and abbreviations (such as GP and EGA) will be used to indicate bibliographical references.

AMSTERDAM, January 1966

F. OORT

Contents

Chapter I PRELIMINARIES

I.1 Group schemes

Let $\underline{C}$ be a category. A group object in $\underline{C}$ is a representable functor

$$\underline{C}^{o} \longrightarrow \underline{Gr}$$

with values in the category of groups (cf. [17] , [12] , EGA $0_{III}.8.2$); i.e. there is given an object $G \in \underline{C}$ such that $\mathrm{Hom}_{\underline{C}}(-,G)$ is "functorially a group". It is clear what is meant by a homomorphism of group objects. If products and a final object (point object) exist in $\underline{C}$, the multiplication morphism and the zero morphism of a group object can be given; in that case the structure of a group object on $G \in \underline{C}$ is given by morphisms

$$s:\ G \Pi G \longrightarrow G$$

$$i:\ G \longrightarrow G \qquad\qquad 0:\ pt \longrightarrow G,$$

which satisfy the usual group axioms: the following diagrams are required to be commutative:

$$\begin{array}{ccc} G\Pi G\Pi G & \xrightarrow{\ s\Pi 1_G\ } & G\Pi G \\ \Big\downarrow{\scriptstyle 1_G\Pi s} & & \Big\downarrow{\scriptstyle s} \\ G\Pi G & \xrightarrow{\ s\ } & G \end{array} \qquad \text{(associative composition),}$$

$$\begin{array}{ccc} G & \xrightarrow{\ (0,1_G)\ } & G\Pi G \\ & {\scriptstyle 1_G}\searrow & \Big\downarrow{\scriptstyle s} \\ & & G \end{array} \qquad \text{(left identity),}$$

$$\begin{array}{ccccc} G & \xrightarrow{\ \Delta=(1,1)\ } & G\Pi G & \xrightarrow{\ (i,1)\ } & G\Pi G \\ & \searrow & & & \Big\downarrow{\scriptstyle s} \\ & & pt & \xrightarrow{\ 0\ } & G \end{array} \qquad \text{(left inverse).}$$

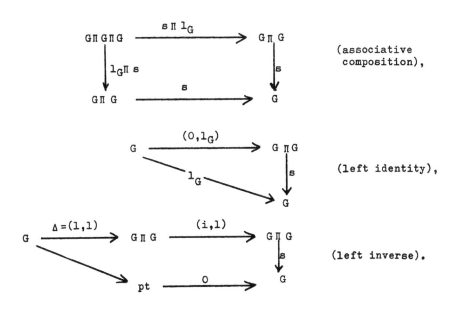

We say that G is a commutative group object if moreover s is a commutative composition.

Let S be a prescheme and let $G \longrightarrow S$ be a seperated morphism. We say that G is a (commutative) group scheme over S if it is a (commutative) group object in the category (Sch/S) of preschemes over S. In this category products (even fibred products) exist, and S is a final object. Hence the structure of a group scheme G over S can be given with the help of S-morphisms

$$s_G : G \times G \longrightarrow G$$

$$i_G : G \longrightarrow G \qquad 0_G : S \longrightarrow G,$$

which satisfy the relations explained above. We denote by $\underline{G}_S$ the category of commutative group schemes of finite type over S. We now give some examples and definitions.

Let S be a prescheme. Consider the functor $(Sch/S) \longrightarrow \underline{Ab}$ into the category of abelian groups defined by

$$T \longmapsto \Gamma\,(T, \underline{O}_T)^+$$

T a prescheme over S (cf FGA, 190-14). This functor is representable; the object representing it we denote by $\underline{\underline{G}}_{aS}$, the additive linear group over S. Below we give a description of this group scheme in case S is affine. We remark that for base change $S' \longrightarrow S$ it holds

$$\underline{\underline{G}}_{aS'} = \underline{\underline{G}}_{aS} \times_S S' .$$

Consider the functor $(Sch/S) \longrightarrow \underline{Ab}$ defined by

$$T \longmapsto \Gamma\,(T, \underline{O}_T)^*$$

T a prescheme over S (if A is a ring, with identity element, we denote by A^* its multiplicative group of units). This functor is representable; the object representing it we denote by $\underline{\underline{G}}_{mS}$, the <u>multiplicative linear group</u> over S. If $S' \longrightarrow S$, it holds

$$\underline{\underline{G}}_{mS'} = \underline{\underline{G}}_{mS} \times_S S'.$$

<u>DEFINITION</u> (cf. FGA, 236-17; cf. [27] , 6.1): Let X be a S-group scheme of finite type. We say that X is an <u>abelian scheme</u> if X has connected fibres, and $X \longrightarrow S$ is smooth (EGA, IV.6.8.1, "lisse"; SGA, 1960.II, "simple") and proper.

<u>REMARKS</u>: "Geometrically connected fibres" (cf. EGA, IV.4.5.2) and "connected fibres" are the same concepts in the case of group schemes (cf. SGAD, VI.2.1.1). An abelian scheme is a commutative group scheme (cf. [27] , 6.1, corollary 6.5). In the case S = Spec(field) an abelian scheme will be called an <u>abelian variety</u> (strictly speaking it is not a variety, as it belongs to a different category). As reduced algebraic schemes and varieties correspond in a natural way (for example compare [39]), we use results concerning algebraic groups in the case of the corresponding algebraic group schemes.

Let A be a ring (commutative, $1 \in A$), and let S = Spec(A). The category of A-algebras and the dual category of the affine S-schemes are equivalent (cf. EGA, I.1.7). Hence affine group schemes over S correspond to cogroup objects in the category of A-algebras (a cogroup object is a group object in the dual category).

<u>DEFINITION</u>: E is called an A-<u>bialgebra</u> if it is a commutative cogroup object in the category of the (associative, commutative) A-algebras ("$\mathrm{RHom}_A(E, -)$ is functorially a group")(the term bialgebra was suggested by M.LAZARD, cf [23] , page 14).

This definition can be made more explicit: the A-algebra structure on E is given by a multiplication m_E, and $1 \in E$ defines $\eta_E : A \longrightarrow E$; the cogroup structure can be given as we did for a group object. Hence an A-bialgebra is an A-module E with A-homomorphisms

$$E \otimes_A E \xrightarrow{m_E} E \qquad E \xrightarrow{s_E} E \otimes_A E$$

$$E \xrightarrow{i_E} E$$

$$A \xrightarrow{\eta_E} E \qquad E \xrightarrow{\varepsilon_E} A,$$

such that m_E and η_E define on E the structure of an A-algebra, and such that s_E, i_E and ε_E are A-algebra homomorphisms defining the structure of a commutative cogroup object on E (use the definition of a group object with all arrows reversed) (we have hesitated on choosing notation; usually the map s_E is called the diagonal map, and it is written as Δ; we avoid this as $\Delta : G \longrightarrow G \times G$, $G = \mathrm{Spec}(E)$, defines m_E, and not s_E. We have choosen the notation ε_E because usually an augmentation is indicated thus).

<u>Explicit examples</u>: $S = \mathrm{Spec}(A)$,

$$E = A[X] \quad , \; s(X) = X \otimes 1 + 1 \otimes X, \; i(X) = -X, \; \varepsilon(X) = 0;$$

it is easily verified that $\underline{\underline{G}}_{aS} = \mathrm{Spec}(E)$.

$$E = A[Y, Y^{-1}] \quad , \; s(Y) = Y \otimes Y, \; i(Y) = Y^{-1}, \; \varepsilon(Y) = 1;$$

in this case $\underline{\underline{G}}_{mS} = \mathrm{Spec}(E)$. Further examples will be given in section 2.

<u>LEMMA</u> (1.1): Let k be a <u>perfect</u> field, and let G be a group scheme over $S = \mathrm{Spec}(k)$. Then G_{red} is a subgroup scheme of G.

<u>PROOF</u>: Clearly the following diagrams are commutative:

$$\begin{array}{ccc} G_{red} & \xrightarrow{i_{red}} & G_{red} \\ \downarrow & & \downarrow \\ G & \xrightarrow{i} & G \end{array} \qquad \qquad \begin{array}{ccc} S & \longrightarrow & G_{red} \\ & \searrow & \downarrow \\ & & G \end{array}$$

As k is perfect, the tensorproduct of k-algebras without nilpotent elements has no nilpotent elements (cf. [5] , 7.5, corollary of proposition 5 and 7.6, corollay 3). Hence

$$(G \times G)_{red} \xrightarrow{\sim} G_{red} \times G_{red}$$

is an isomorphism. Thus we obtain a commutative diagram

$$\begin{array}{ccccc} & \overset{\sim}{\nearrow} G_{red} \times G_{red} & \dashrightarrow & & \\ (G \times G)_{red} & & \xrightarrow{m_{red}} & & G_{red}\,, \\ \downarrow & & & & \downarrow \\ G \times G & & \xrightarrow{m} & & G \end{array}$$

and the lemma is proved.

We now define the Frobenius homomorphism for group schemes in characteristic p different from zero (cf. GP, 1.1; cf. [38] , III.1).

Let p be a prime number and let A be a ring (commutative, $1 \in A$) of characteristic p (i.e. $p.1 = 0$). We define $F: A \longrightarrow A$ by

$$F(a) = a^p, \quad a \in A;$$

this is a ringhomomorphism. Let ${}^aF = \phi : \mathrm{Spec}(A) = S \longrightarrow S$ be the induced morphism. Let X be a prescheme over S. Then all local rings of X are of characteristic p, hence there exists a morphism from X into itself which is obtained by raising to the p-th power in each stalk of the structure sheaf of X. We obtain a commutative diagram

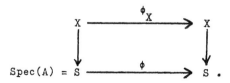

$$\begin{array}{ccc} X & \xrightarrow{\phi_X} & X \\ \downarrow & & \downarrow \\ \mathrm{Spec}(A) = S & \xrightarrow{\phi} & S\,. \end{array}$$

We define $\theta(X)$ by the cartesian square

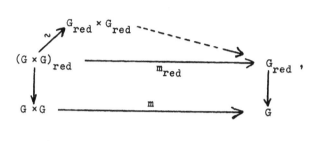

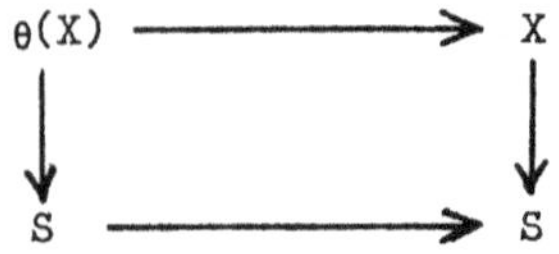

and we define $F: X \longrightarrow \theta(X)$ by the commutativity of the diagram

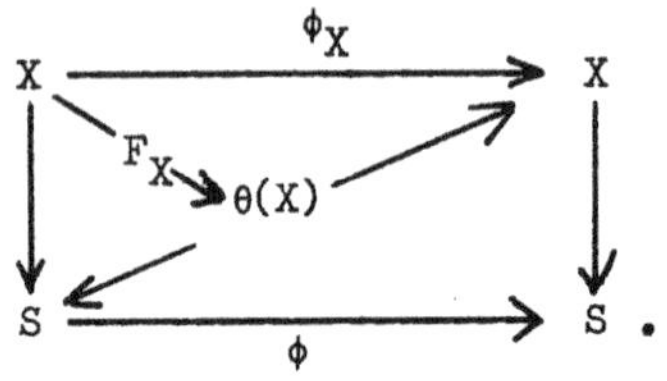

If X is a S-group scheme, $\theta(X)$ is a S-group scheme and F_X is a homomorphism. If X comes from a scheme over the prime field, X and $\theta(X)$ coincide (and we identify them in that case; e.g. $\underline{\underline{G}}_a$, $\underline{\underline{G}}_m$). By induction we define $\theta^n(X)$ and F^n_X for any positive integer n; θ^n is a functor, and F^n is a morphism of functors for each n.

Let G be a group scheme over S. For a positive integer n we write

$$\underline{\underline{I}}^n(G) = \mathrm{Ker}(F^n : G \longrightarrow \theta^n G).$$

This is a subgroup scheme of G; it is defined by the cartesian square

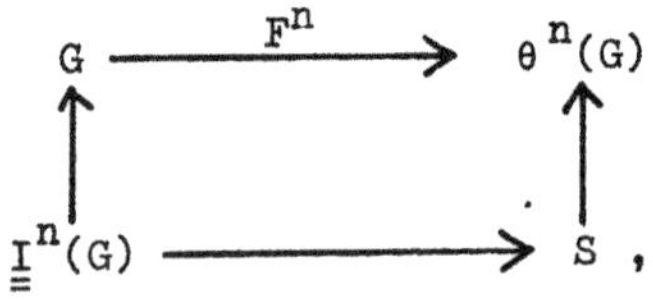

where the right-hand vertical arrow is the zero section of $\theta^n(G)$. We are going to describe the group scheme $\underline{\underline{I}}^n(G)$. We remark that for each n, $\underline{\underline{I}}^n$ is a functor. The zero section $0_G: S \longrightarrow G$ identifies S with a closed subscheme of G (cf. EGA, I.5.3.11). Hence S is defined by a sheaf of ideals $\underline{I} \subset \underline{O}_G$:

$$0 \longrightarrow \underline{I} \longrightarrow \underline{O}_G \longrightarrow \underline{O}_S \longrightarrow 0.$$

<u>LEMMA</u> (1.2): For any positive integer, $\underline{\underline{I}}^n(G)$ is defined by the sheaf of ideals $\underline{I}^{(p^n)} \subset \underline{O}_G$.

[<u>notation</u>: let I be an ideal in a ring R; we denote by $I^{(p^n)}$ the ideal generated by all elements x^{p^n}, $x \in I$; N.B. in general I^{p^n} and $I^{(p^n)}$ are different.]

[<u>special case</u>: $S = \mathrm{Spec}(k)$, k a field; then $\underline{I} = \underline{m}_{G,e}$, the maximal ideal of the closed point $0_G = e \in G_k$, and $\underline{\underline{I}}^n(G) = \mathrm{Spec}(\underline{O}_{G,e}/\underline{m}_{G,e}^{(p^n)})$.]

<u>PROOF</u>: Let $s \in S \subset G$. Consider the commutative diagram

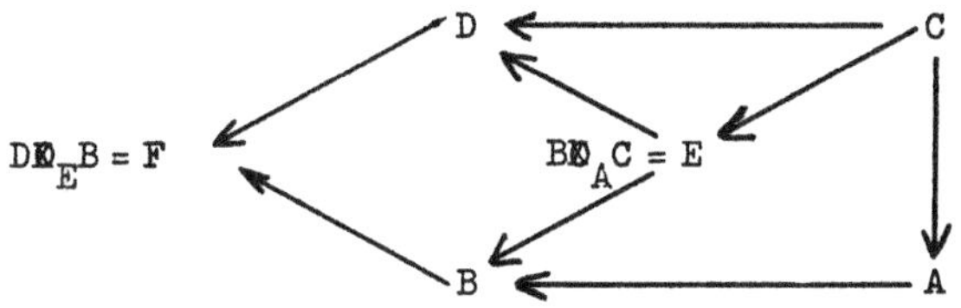

($D = \underline{O}_{G,s} = C$, $B = \underline{O}_{S,s} = A$, the horizontal arrows stand for exponentiation by p^n, $A \subset C$ and $C \longrightarrow A$ is an A-algebra homomorphism, $B \subset E$ and $E \longrightarrow B$ is an B-algebra homomorphism). The ideal $I = \underline{I}_s$ fits into the exact sequence of C-modules:

$$0 \longrightarrow I \longrightarrow C \longrightarrow A \longrightarrow 0.$$

Hence the sequence of E-modules

$$B \otimes_A I \longrightarrow B \otimes_A C \longrightarrow B \otimes_A A \longrightarrow 0$$

is exact. As $B \otimes_A C = E$ and $B \otimes_A A \cong B$, we obtain an exact, commutative diagram

$$\begin{array}{ccccccc} & \nearrow & E \otimes_E D & \longrightarrow & B \otimes_E D & \longrightarrow & 0 \\ (B \otimes_A I) \otimes_E D & & \updownarrow \cong & & \updownarrow = & & \\ & \searrow & D & \xrightarrow{\ f\ } & F & \longrightarrow & 0. \end{array}$$

Let $J = \mathrm{Ker}(f)$. Choose $(b \otimes a) \otimes d \in (B \otimes_A I) \otimes_E D$. Its image in D equals $bp^n ad \in I^{(p^n)} \subset \underline{O}_{G,s}$. Hence $J \subset I^{(p^n)}$. Let $x \in I^{(p^n)}$. Then we can write

$$x = \sum x_i^{p^n} d_i \qquad x_i \in I, \quad d_i \in D.$$

This element is the image of $\sum (x_i \otimes 1) \otimes d_i$; hence $J \subset I^{(p^n)}$. This proves

$$F = D/I^{(p^n)} .$$

Thus $\underline{\underline{I}}^n(G)$ is defined by the sheaf of ideals $\underline{I}^{(p^n)}$, which proves the lemma.

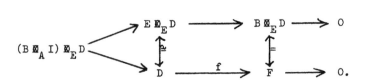

I.2 Finite group schemes

We start with some easy results in commutative algebra. Let A be a ring (commutative, $1 \in A$); we denote by $\underline{Ab}_A$ the category of A-modules. For $E \in \underline{Ab}_A$ we define

$$E^D = \mathrm{Hom}_A(E,A) \in \underline{Ab}_A$$

(A being commutative, we do not distinguish left and right modules). For $E,F \in \underline{Ab}_A$ and $f \in \mathrm{Hom}_A(E,F)$ we write

$$\mathrm{Hom}_A(f,A) = f^D : F^D \longrightarrow E^D ;$$

we define

$$\Phi = \Phi_A(E,F): \; E \otimes_A F \longrightarrow \mathrm{Hom}_A(E^D,F)$$

by

$$(\Phi\,(a \otimes b))(f) = f(a)b, \qquad a \in E,\ b \in B,\ f \in E^D.$$

<u>LEMMA</u> (2.1): $\Phi_{-}(-,-)$ is a morphism of trifunctors.

This is proved by a direct verification.

<u>LEMMA</u> (2.2): If $E \in \underline{Ab}_A$ is finitely generated and projective over A, $\Phi(E,B)$ is an isomorphism for every $B \in \underline{Ab}_A$.

<u>PROOF</u>: Assume first E to be finitely generated and free:

$$E \cong A^n \quad \text{for some integer } n.$$

The following isomorphisms are well-known:

$$E \otimes_A B \cong A^n \otimes_A B \cong (A \otimes_A B)^n \cong B^n$$

and

$$\mathrm{Hom}_A(E^D,B) \cong \mathrm{Hom}_A(A^n,B) \cong (\mathrm{Hom}_A(A,B))^n \cong B^n.$$

The following diagram commutes

$$\begin{array}{ccc} B^n & \xrightarrow{\;\sim\;} & \mathrm{Hom}_A(E^D,B), \\ \uparrow \wr & \nearrow_{\Phi} & \\ E \otimes_A B & & \end{array}$$

hence the case $E = A^n$ is proved. As a finitely generated, projective A-module is a direct summand of A^n for some n, the lemma follows.

For $E \in \underline{Ab}_A$, we define

$$\kappa = \kappa_E : E \longrightarrow E^{DD}$$

by

$$(\kappa(a))(f) = f(a), \quad a \in E, \ f \in E^D.$$

Note that κ_E is functorially in E.

COROLLARY (2.3): Let E be finitely generated and projective over A. Then

$$\kappa = \kappa_E : E \xrightarrow{\sim} E^{DD}$$

is an isomorphism.

PROOF:

$$\begin{array}{ccc} E & \xrightarrow{\kappa} & E^{DD} \\ \uparrow \wr\!\!\!\| & & \| \\ E \otimes_A A & \xrightarrow[\Phi]{\sim} & \mathrm{Hom}_A(E^D, A). \end{array}$$

QED

LEMMA (2.4): If E is finitely generated and projective over A, the same holds for E^D.

PROOF: If $E \oplus F = A^n$, $E^D \oplus F^D = (A^n)^D \cong A^n$.

QED

COROLLARY (2.5): Let $E \in \underline{Ab}_A$ be finitely generated and projective. Then

$$E^D \otimes_A E^D \xrightarrow{\sim} (E \otimes_A E)^D$$

is an isomorphism.

PROOF: $$(E^D \otimes_A E^D \xrightarrow[\Phi]{\sim} \mathrm{Hom}_A(E^{DD}, E^D) \xrightarrow[\kappa^*]{\sim} \mathrm{Hom}_A(E, E^D))$$

is an isomorphism by (2.4), (2.2) and (2.3); also

$$\mathrm{Hom}(E, \mathrm{Hom}(E, A)) \xrightarrow{\sim} \mathrm{Hom}(E \otimes E, A) = (E \otimes E)^D$$

is an isomorphism (cf. [8] , II.5, proposition 5.2).

QED

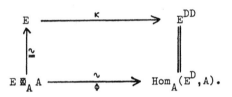

Alternative proof: E is a direct summand of A^n, and E^{DD} is a direct summand of $(A^n)^{DD} = A^n$, etc.

Under the conditions of (2.5) we identify $E^D \otimes_A E^D$ and $(E \otimes_A E)^D$.

DEFINITION: Let S be a prescheme. We say that N is a finite group scheme over S, if it is a (commutative) group scheme over S, such that the structural morphism $N \longrightarrow S$ is finite (EGA, II.6.1.1; i.e. we can find an open covering $\{S_\alpha\}$ of S such that $N \times_S S_\alpha = \mathrm{Spec}(E_\alpha)$, where E_α is finitely generated, as a module, over $\Gamma(S_\alpha, \underline{O}_S)$).

We remark that every fibre of a finite morphism is finite. A more correct terminology would be: "N is a group scheme, finite over S".

Let $N = \mathrm{Spec}(E)$ be a finite commutative $S = \mathrm{Spec}(A)$— group scheme. Suppose E to be a projective (finitely generated) A-module. We define:

$$s_{E^D} = (m_E)^D : E^D \longrightarrow E^D \otimes E^D \xrightarrow{\sim} (E \otimes E)^D,$$

$$m_{E^D} = (s_E)^D : E^D \otimes E^D \xrightarrow{\sim} (E \otimes E)^D \longrightarrow E^D,$$

$$i_{E^D} = (i_E)^D : E^D \longrightarrow E^D,$$

$$\varepsilon_{E^D} = (\eta_E)^D : E^D \longrightarrow A^D \cong A,$$

$$\eta_{E^D} = (\varepsilon_E)^D : A \cong A^D \longrightarrow E^D.$$

PROPOSITION (2.6): Let E be a finitely generated, projective, commutative A-bialgebra. Then $E^D = \mathrm{Hom}_A(E,A)$ is an A-bialgebra; D is a contravariant functor and DD is isomorphic with the identity functor.

PROOF: Straightforward verification proves that m_{E^D} and η_{E^D} equip E^D with the structure of a (commutative, associative) A-algebra with $\eta_{E^D}(1) = 1 \in E^D$; moreover the remaining three maps are A-algebra homomorphisms, which satisfy the conditions for a cogroup object. It is clear that an A-bialgebra homomorphism $\phi : E \longrightarrow F$ defines an A-bialgebra homomorphism $\phi^D : F^D \longrightarrow E^D$; $DD \cong \mathrm{id}$ follows from (2.3).

QED

REMARK: If E is a commutative A-bialgebra (not necessarily projective), E^D is equiped with the structure of a commutative A-algebra with augmentation.

LEMMA (2.7) (cf. [4] , page 20, proposition A.1): Let $E \in \underline{Ab}_A$; this module is finitely generated and projective over A if and only if the composition

$$\rho_E = (E^D \otimes E \xrightarrow{\Phi} \mathrm{Hom}(E^{DD},E) \xrightarrow{\kappa^*} \mathrm{Hom}(E,E))$$

is surjective (and in that case it is an isomorphism).

PROOF: Suppose E to be finitely generated and projective over A. By (2.4) and (2.2) we conclude that $\Phi = \Phi_A(E^D,E)$ is an isomorphism, and by (2.3) it follows that κ is an isomorphism.

Suppose conversely that $1_E \in \mathrm{Im}(\kappa^* \Phi)$; in that case there exist $f_i \in E^D$, $x_i \in E$ such that

$$\rho_E(\Sigma\, f_i \otimes x_i) = 1_E \,.$$

For $x \in E$ it holds

$$x = (\rho_E(\Sigma f_i \otimes x_i))(x) = \Sigma\, f_i(x)\, x_i \,;$$

hence the set $\{x_i\}$ generates E. Thus we can choose a finitely generated free A-module F and an epimorphism $F \longrightarrow E$. Then we obtain a commutative diagram (use 2.1):

$$\begin{array}{ccccc} E^D \otimes F & \longrightarrow & E^D \otimes E & \longrightarrow & 0 \\ \downarrow & & \downarrow{\rho_E} & & \\ \mathrm{Hom}(E,F) & \longrightarrow & \mathrm{Hom}(E,E) & & \end{array}$$

with exact upper row (the tensorproduct is right exact). Hence E is a direct summand of F.

QED

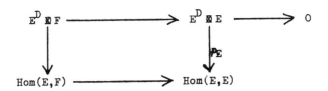

COROLLARY (2.8): Let A be a ring and $E \in \underline{Ab}_A$. If E has a finite presentation and E is flat over A, E is A-projective.

PROOF: If E has a finite presentation, there exist finitely generated free A-modules L and F and an exact sequence

$$L \longrightarrow F \longrightarrow E \longrightarrow 0.$$

As E is flat, there results a commutative exact diagram

$$\begin{array}{ccccccc}
0 & \longrightarrow & E^D \otimes E & \longrightarrow & F^D \otimes E & \longrightarrow & L^D \otimes E \\
 & & \downarrow{\scriptstyle \rho_E} & & \downarrow{\scriptstyle \kappa\Phi} & & \downarrow{\scriptstyle \kappa\Phi} \\
0 & \longrightarrow & \mathrm{Hom}(E,E) & \longrightarrow & \mathrm{Hom}(F,E) & \longrightarrow & \mathrm{Hom}(L,E).
\end{array}$$

As F and L are finitely generated and free, the second and third vertical arrows are isomorphisms. Hence ρ_E is surjective, and we conclude by (2.7).

QED

REMARK: For a different proof, compare the suggestions in [6], page 64, exercise 15 and page 66, exercise 23c.

NOTATION: Let S be a prescheme. We denote by $\underline{N}_S$ the category of commutative, finite, flat S-group schemes. If $S = \mathrm{Spec}(A)$, we write $\underline{A}_A = \underline{N}_S^o$, the category of commutative, finitely generated, flat A--bialgebras.

The result we obtained can be summarized as follows:

PROPOSITION (2.9): Let S be a locally noetherian prescheme. We have defined a contravariant functor

$$D: \underline{N}_S^o \longrightarrow \underline{N}_S,$$

which is a duality:

$$\kappa : \mathrm{id} \xrightarrow{\sim} DD$$

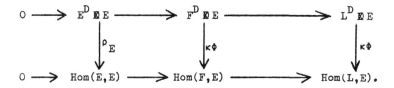

is an isomorphism of functors. The functor D commutes with base change, i.e. if $T \longrightarrow S$ is a morphism of preschemes, and $N \in \underline{N}_S$, the natural homomorphism

$$(N_T)^D \xrightarrow{\sim} (N^D)_T$$

is an isomorphism.

PROOF: It suffices to prove these facts over an affine base scheme. If $N = \mathrm{Spec}(E) \in \underline{N}_S$, E is a projective module over $A = \Gamma(S, \underline{O}_S)$ by (2.8). Thus D is defined by (2.6), and κ is an isomorphism. Consider the homomorphisms

$$\mathrm{Hom}_A(E,A) \otimes_A B \xrightarrow{\sim} \mathrm{Hom}_A(E,B) \xrightarrow{\sim} \mathrm{Hom}_B(E \otimes_A B, B).$$

The second is an isomorphism (cf. [8], page 30, line 3). The first homomorphism is an isomorphism if E is finitely generated and free over A, hence it is an isomorphism if E is finitely generated and projective over A.

QED

We identify $(N_T)^D$ and $(N^D)_T$ in the case just described, and we write N_T^D. The duality D is well-known in the case of a base field (cf. [16], page I-9, lemma 4; cf. [11], section 14); we say that N^D is the linear dual of N. In the case of a perfect base field, the structure of $\underline{N}_S$ was determined by GABRIEL (cf. [16]). We now try to give a brief review of part of his methods.

For the rest of this section we fix the following: k is a perfect field of characteristic p different from zero, $S = \mathrm{Spec}(k)$, $\underline{N} = \underline{N}_S$, $\underline{A} = \underline{A}_k$ the category of commutative finitely generated k-bialgebras (they are automatically flat over k).

[In case of characteristic zero there are no difficulties: all algebraic group schemes are reduced by a result of CARTIER, cf. [11], page 109, cf. [26], page 25.1, theorem 1; if moreover k is algebraically closed,

$\underline{N}_k$ is equivalent with the category of finite abelian groups.]

We write $\underline{A}_{loc}$ for the full subcategory of $\underline{A}$ consisting of all $E \in \underline{A}$ which are local rings, and $\underline{A}_{red}$ for the full subcategory of all $F \in \underline{A}$ with $F = F_{red}$ (i.e. the ring F has no nilpotent elements).

LEMMA (2.10): Let $E \in \underline{A}_{loc}$ and $F \in \underline{A}_{red}$. Then

$$\mathrm{Hom}_{\underline{A}}(E,F) = 0 \quad \text{and} \quad \mathrm{Hom}_{\underline{A}}(F,E) = 0.$$

PROOF: A ring homomorphism $f: E \longrightarrow F$ contains the maximal ideal of E, as F has no nilpotent elements. Hence the first statement follows.

Next we note the following facts:

a) The only idempotent elements in a local ring are 0 and 1 (suppose $x^2 = x$; if $x \in \underline{m}_E$, then $x-1 \notin \underline{m}_E$, hence this element has an inverse and $(x-1)x = 0$ implies $x = 0$; if $x \notin \underline{m}_E$, this element has an inverse and $(x-1)x = 0$ implies $x-1 = 0$).

b) Let $F \in \underline{A}_{red}$. As F is a reduced artinian ring, it can be written as a product of fields (theorem of Dedekind, cf. [40], § 143):

$$F = K_1 \times \ldots \times K_n.$$

Let $e_1, \ldots, e_n$ be the corresponding idempotents of F. As $\varepsilon_F \eta_F = 1_k$, we can choose the decomposition in such a way that $\varepsilon_F(e_1) = 1$. Then $K_1 = k$, and as $e_1 e_i = 0$ for $2 \leq i \leq n$, $\varepsilon_F(e_i) = 0$.

If $g \in \mathrm{Hom}_{\underline{A}}(F,E)$, $\varepsilon_F = \varepsilon_E g$, thus $g(e_1) = 1$ and $g(e_i) = 0$ for $2 \leq i \leq n$. This proves $g = \eta_E \varepsilon_F$, thus $\mathrm{Hom}_{\underline{A}}(F,E) = 0$, and the lemma is proved.

We prove some facts about Cohen-subrings which are to be found in the literature (for example cf. [30], section 31).

LEMMA (2.11): Let E be a local artinian ring with maximal ideal $\underline{m}_E$

and perfect residue field $K_E = E/\underline{m}_E$. Suppose E is of characteristic $p \neq 0$. Then the residue field can be lifted (i.e. E has a subfield $K_E' \subset E$ which is isomorphic with K_E under $E \longrightarrow K_E$).

PROOF: As E is of characteristic p (i.e. in E holds $p.1 = 0$), for $x, y \in E$ we obtain $(x+y)^p = x^p + y^p$. We choose a number n such that $x^{p^n} = 0$ for all $x \in \underline{m}_E$. Let $a \in K_E$; as K_E is perfect, there exists $b \in K_E$ such that $b^{p^n} = a$; choose $c \in E$ such that $c \text{ mod.} \underline{m}_E = b$; we defined $\sigma(a) = c^{p^n}$. In this way we defined a homomorphism

$$\sigma = \sigma_E : K_E \longrightarrow E \quad \text{such that} \quad (K_E \xrightarrow{\sigma} E \longrightarrow K_E) = 1_{K_E},$$

and the lemma is proved.

REMARK: The subring $\sigma(K_E) = K_E' \subset E$ is called the Cohen subring of E; its construction can be given in the case of an arbitrary complete local ring.

LEMMA (2.12): Let k be a perfect field of characteristic $p \neq 0$. Let E and F be local, finitely generated k-algebras, and let $f: E \longrightarrow F$ be a (local) homomorphism of k-algebras; let $f': K_E \longrightarrow K_F$ be the induced homomorphism. Then the diagram

$$\begin{array}{ccc} E & \xleftarrow{\sigma_E} & K_E \\ {\scriptstyle f}\downarrow & & \downarrow{\scriptstyle f'} \\ F & \xleftarrow{\sigma_F} & K_F \end{array}$$

is commutative ("the construction of the Cohen subrings commutes with ringhomomorphisms").

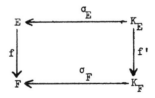

PROOF: E and F contain k and K_E and K_F are finite extensions of k (hence they are perfect); thus the conditions of (2.11) are satisfied, thus the sections σ_E, σ_F exist. As $(\underline{m}_E \longrightarrow F \longrightarrow K_F) = 0$, the existence of f' follows. We have to prove that

$$f(K_E') \subset K_F' = \sigma_F(K_F)$$

(and then the result follows easily). We write $L = f(K_E') \cap K_F'$. This field is an extension of k, and even a finite extension, hence it is a perfect field. Let $x \in f(K_E')$; then

$$x = a + b, \qquad a \in K_F', \quad b \in \underline{m}_F \,;$$

we choose $q = p^n$ such that $b^q = 0$. Then $x^q = (a+b)^q = a^q \in L$, hence (L being perfect) $x \in L$. This proves the lemma.

Let $E \in \underline{A}$. As k is a perfect field, $E_{red} \in \underline{A}$ (cf. 1.1), and $E \longrightarrow E_{red}$ is a homomorphism of bialgebras.

<u>COROLLARY</u> (2.13): E_{red} is a direct summand of E in the category $\underline{A}$.

<u>PROOF</u>: E can be written as a product of local, finitely generated k-algebras. Hence by (2.11) we can consider E_{red} as a subalgebra of E (the "Cohen subalgebra" of E). Note that $E_{red} \otimes E_{red} = (E \otimes E)_{red}$. From (2.12) commutativity of the diagram

$$\begin{array}{ccc} E & \xleftarrow{\ \sigma_E\ } & E_{red} = F \\ \downarrow{\scriptstyle s_E} & & \downarrow{\scriptstyle s_F} \\ E \otimes E & \xleftarrow{\ \sigma_{E\otimes E}\ } & F \otimes F \end{array}$$

follows; hence σ_E is a homomorphism in $\underline{A}$, and the corollary is proved.

<u>PROPOSITION</u> (2.14): Let k be a perfect field. Every $E \in \underline{A}$ can be written in a unique way in the form

$$E = E_{red} \oplus E_{loc}, \quad E_{red} \in \underline{A}_{red} \quad \text{and} \quad E_{loc} \in \underline{A}_{loc}$$

($\oplus$ stands for the direct sum in $\underline{A}$), and

$$\mathrm{Hom}_{\underline{A}}(E,F) = \mathrm{Hom}_{\underline{A}}(E_{loc},F_{loc}) \times \mathrm{Hom}_{\underline{A}}(E_{red},F_{red}).$$

PROOF: Consider the diagram

$$\begin{array}{ccc} E & \xleftarrow{\ \sigma_E\ } & E_{red} = F \\ \downarrow & & \downarrow \varepsilon_F \\ E_{loc} = E \otimes_F k & \longleftarrow & k \end{array}$$

which defines E_{loc}. As E is a finitely generated k-algebra we can write $E = E_1 \times \dots \times E_n$ as a product of local k-algebras; suppose $\varepsilon_E(e_1) = 1$ and $\varepsilon_E(e_i) = 0$, $2 \leq i \leq n$ (where $e_1,\dots,e_n$ are the corresponding idempotents). Clearly $E \otimes_F k = E_1 = E_{loc}$, hence $E_{loc} \in \underline{A}_{loc}$. As $E_{loc} = \mathrm{Coker}(\sigma_E)$ it follows by (2.13) that $E = E_{red} \oplus E_{loc}$. The second half of the proposition follows from (2.10); thus uniqueness of the decomposition follows.

QED

The linear dual E^D of E can be decomposed in the same way. In this way the category $\underline{A}$ is split up into four categories:

$\underline{A}_{ll}$	the category of all	$E \in \underline{A}_{loc}$	with	$E^D \in \underline{A}_{loc}$
$\underline{A}_{lr}$	- - - -	$E \in \underline{A}_{loc}$	-	$E^D \in \underline{A}_{red}$
$\underline{A}_{rl}$	- - - -	$E \in \underline{A}_{red}$	-	$E^D \in \underline{A}_{loc}$
$\underline{A}_{rr}$	- - - -	$E \in \underline{A}_{red}$	-	$E^D \in \underline{A}_{red}$.

For every $E \in \underline{A}$ we can write:

$$E = E_{loc} \oplus E_{red} = (E_{11} \oplus E_{1r}) \oplus (E_{r1} \oplus E_{rr}).$$

It can be proved that each of these categories (and hence $\underline{A}$ itself) is abelian (cf. GABRIEL, [16], page I-15, theorem 1; it is a special case of a result of GROTHENDIECK, cf. FGA, page 212-17, corollary 7.4). The functor

$$D: \underline{A}^{o} \longrightarrow \underline{A}$$

is exact.

If $E \in \underline{A}_{loc}$, the Frobenius homomorphism of $N = \mathrm{Spec}(E)$ is nilpotent; in this way E_{loc} and E_{red} can be characterized, and the same can be done for E^{D} (cf. [23], I.3; remark that in this paper on page 19, proposition 1.5 the categories occuring in (b) and (c) must be interchanged).

For $\underline{N}$ we use the analogous notations: $\mathrm{Spec}(E) = N \in \underline{N}_{loc}$ if $E \in \underline{A}_{loc}$, $\underline{N} = \underline{N}_{loc} \oplus \underline{N}_{red}$, etc.

We recall that a simple object of an abelian category is an object different from zero with no proper subobjects different from zero (cf. [17], page 337); now we describe simple objects in $\underline{N}$.

In the category $\underline{N}_{11}$ there is only one simple object (cf. [16], III.1, lemma 1; cf. [38], III.2, proposition 11; compare section 16). It will be denoted by α_p:

$$\alpha_p \in \underline{N}_{11}, \quad E = \Gamma(\underline{O}_{\alpha_p}) \in \underline{A}_{11},$$

$$E = k[X]/(X^p) \cong k[\tau],$$

$$s(\tau) = \tau \otimes 1 + 1 \otimes \tau,$$

$$i(\tau) = -\tau, \quad \varepsilon(\tau) = 0.$$

The sequence

$$0 \longrightarrow \alpha_p \longrightarrow \underline{\underline{G}}_a \xrightarrow{F} \underline{\underline{G}}_a \longrightarrow 0$$

is exact (no hypothesis on k is needed).

We introduce the following finite group schemes ("roots of unity")

$$\mu_q = \operatorname{Ker}(\underline{\underline{G}}_m \xrightarrow{q.\mathrm{id}} \underline{\underline{G}}_m) \qquad q \text{ a positive integer.}$$

If $(q,p)=1$, $\mu_q \in \underline{N}_{red}$ and $\mathrm{Mor}_S(S, \mu_q) = \mathbb{Z}/q\,\mathbb{Z}$ (the algebraic group corresponding with μ_q in case $(q,p)=1$ was denoted by $\mathbb{Z}/q\,\mathbb{Z}$ in GP). The group scheme μ_p can be described as follows:

$$\mu_p \in \underline{N}_{loc}, \quad E = \Gamma(\underline{O}_{\mu_p}),$$
$$E = k[Y]/(Y^p - 1) = k[\rho] ,$$
$$s(\rho) = \rho \otimes \rho ,$$
$$i(\rho) = \rho^{-1} , \quad \varepsilon(\rho) = 1.$$

We define ν_{p^n} by the exact sequence

$$0 \longrightarrow \nu_{p^n} \longrightarrow \underline{\underline{G}}_a \xrightarrow{\mathrm{id} - F^n} \underline{\underline{G}}_a \longrightarrow 0;$$

clearly $\nu_{p^n} \in \underline{N}_{red}$ and $\mathrm{Mor}_S(S, \nu_{p^n}) = \mathbb{Z}/p^n\,\mathbb{Z}$.

<u>LEMMA</u> (2.15):

$$\mu_q^D \cong \mu_q , \qquad \text{if } (q,p)=1;$$

$$\mu_p^D \cong \nu_p , \text{and hence} \qquad \nu_p^D \cong \mu_p .$$

<u>PROOF</u>: Explicit calculation or application of the Cartier-Shatz formula (cf. [11], page 107, (22); cf. section 16).

QED

LEMMA (2.16): Let k be an algebraically closed field. The functor

$$\underline{N}_{red} \longrightarrow \underline{FAb}$$

which assigns to every $N \in \underline{N}_{red}$ the finite abelian group $Mor_S(S,N) \in \underline{FAb}$ is an equivalence of categories.

[In case k is a perfect field, not necessarily algebraically closed, Galois theory has to be used: cf. [16], page I-11.]

PROOF: If E is a finitely generated k-algebra, $RHom_k(E,k)$ is a finite set (and, as k is algebraically closed, $E_{red} \cong k \oplus \ldots \oplus k$ by the theorem of Dedekind). If V is a finite set, Map(V,k) is in a natural way a reduced, finitely generated k-algebra. Thus we have defined contravariant functors

$$E \longmapsto RHom_k(E,k)$$
$$V \longmapsto Map(V,k)$$

between the category $\underline{k}$ of finitely generated, reduced k-algebras and the category $\underline{FEns}$ of finite sets. The maps

$$\phi_E: E \xrightarrow{\sim} Map(RHom_k(E,k),k) \qquad \phi_E(x)(g) = g(x),$$
$$\psi_V: V \xrightarrow{\sim} RHom_k(Map(V,k),k) \qquad \psi_V(v)(f) = f(v),$$

are bijective (as k is algebraically closed), and the first one is a homomorphism. Thus we have established an equivalence of the categories $\underline{k}^o$ and $\underline{FEns}$. Thus their categories of group objects $\underline{N}_{red}$ and $\underline{FAb}$ are equivalent, and the proof of the lemma is concluded.

It follows, in the case k is algebraically closed, that ν_p and μ_q, where q is a prime number different from p, are the only simple objects of $\underline{N}_{red}$. Using (2.15) it follows that $\underline{N}_{rl}$ is equivalent with the category of p-primary finite abelian groups, and that $\underline{N}_{rl}$ is equivalent with the category of finite abelian groups with order prime to p. Hence the only simple object of $\underline{N}_{lr}$ (in case k is algebraically closed) is $\nu_p^D = \mu_p$.

I.3 Yoneda extensions

In this section we recall some facts about higher extension groups which are to be found in the literature (cf. [42,43] and [7]).

Let $\underline{C}$ be an abelian category. Following YONEDA one can define the functors $E^n_{\underline{C}}(-,-) = E^n(-,-)$ of n-fold extension groups (without the use of injective or projective objects in $\underline{C}$). For $A,B \in \underline{C}$ an element of $E^n(A,B)$ is the equivalence class containing an exact sequence

$$0 \longrightarrow B \longrightarrow X_n \longrightarrow \cdots \longrightarrow X_1 \longrightarrow A \longrightarrow 0,$$

while the equivalence relation between such exact sequences is the strictest one making equivalent the two rows of the commutative diagram

$$\begin{array}{ccccccccccccc} 0 & \longrightarrow & B & \longrightarrow & X_n & \longrightarrow & \cdots & \longrightarrow & X_1 & \longrightarrow & A & \longrightarrow & 0 \\ & & \| & & \downarrow & & & & \downarrow & & \| & & \\ 0 & \longrightarrow & B & \longrightarrow & Y_n & \longrightarrow & \cdots & \longrightarrow & Y_1 & \longrightarrow & A & \longrightarrow & 0. \end{array}$$

This set $E^n(A,B)$ of equivalence classes can be equiped with the structure of a commutative group (in case $n = 1$: Baer-multiplication, cf. [8], XIV.1, cf. GA. VII.1). If

$$0 \longrightarrow A_1 \longrightarrow A_2 \longrightarrow A_3 \longrightarrow 0$$

is an exact sequence in $\underline{C}$, the functors $E^n(-,B)$ (we write $E^0(-,-) = = \mathrm{Hom}(-,-)$) yield the usual exact cohomology sequence, and analogously for $B_2/B_1 = B_3$ and $E^n(A,-)$ (one can even show that $E^\bullet(-,-)$ is a universal cohomology functor in the sense of [18]). As usual we write $\mathrm{Ext}(-,-) = E^1(-,-)$. For $A,B,C \in \underline{C}$ one defines the Yoneda-pairing:

$$E^n(B,A) \times E^m(C,B) \longrightarrow E^{n+m}(C,A) \quad n \geq 0,\ m \geq 0.$$

For $n > 0$ and $m > 0$ or $n = 0 = m$ the definitions are obvious. For $n = 0$ and $m > 0$ (and dually for $n > 0$ and $m = 0$), the product $f.\xi \in E^n(C,A)$ is defined by the lower exact row of the commutative diagram:

$$\begin{array}{ccccccccccccc}
\xi & 0 & \longrightarrow & B & \longrightarrow & X_m & \longrightarrow \cdots \longrightarrow & X_1 & \longrightarrow & C & \longrightarrow & 0 \\
 & & & \downarrow f & & \downarrow & & \downarrow & & \| & & \\
f.\xi & 0 & \longrightarrow & A & \longrightarrow & Y_m & \longrightarrow \cdots \longrightarrow & Y_1 & \longrightarrow & C & \longrightarrow & 0
\end{array}$$

(cf. GA, VII.1). The Yoneda-product is bilinear and associative. Associativity of

$$E^m(B,A) \times E^n(C,B) \times E^q(D,C) \longrightarrow E^{m+n+q}(D,A)$$

is trivial except for the case $m = 0$, $n = 1$, $q = 0$ (cf. GA, VII.1; cf. [33], page 230, lemma).

LEMMA (3.1): For $A,B \in \underline{C}$ and for any natural number m and any natural number q,

$$(m.1_A)^* = (m.1_B)_* : E^q(A,B) \longrightarrow E^q(A,B)$$

equals to multiplication by m.

This follows immediately from the fact that the Yoneda multiplication is linear.

If in an abelian category sufficiently many injective or projective objects are available (i.e. any object can be imbeddedin an injective object, respectively is quotient of a projective object), the derived functors $\mathrm{Ext}^n(-,-)$ of $\mathrm{Hom}(-,-)$ can be defined, and they are isomorphic (in two ways, cf. GP, 3.5, proposition 10) with $E^n(-,-)$.

If $\underline{C}$ is an abelian category we say that the homological dimension of $\underline{C}$ equals n if n+1 is the smallest integer such that the functor $E^{n+1}(-,-)$ is zero (if $\underline{C}$ is the category of modules over a ring, the homological dimension of $\underline{C}$, and the global dimension of that ring are the same, cf. [8], page 111.).

I.4 Pro-categories

Let $\underline{C}$ be a category. Following GROTHENDIECK (cf. FGA, page 195-03) one can define the category Pro($\underline{C}$): the objects are projective systems of objects of $\underline{C}$ and

$$\mathrm{ProHom}(\{X_i\},\{Y_j\}) = \varprojlim_j \varinjlim_i \mathrm{Hom}_{\underline{C}}(X_i,Y_j)$$

(it must be made clear that the systems are indexed by sets, belonging to a universe, etc.). In a natural way $\underline{C}$ is a full subcategory of Pro($\underline{C}$), namely $\mathrm{Hom}_{\underline{C}}(X,Y) = \mathrm{ProHom}(\{X\},\{Y\})$.

Let $\underline{C}$ be an abelian category. Consider the category Sex($\underline{C}$,$\underline{Ab}$) consisting of left exact, covariant functors from $\underline{C}$ into the category $\underline{Ab}$ of abelian groups, with morphisms of functors as morphisms. This category is an abelian one with exact inductive limits (cf. GABRIEL, [17], page 353, proposition 5). There is a natural functor

$$\underline{C} \longrightarrow (\mathrm{Sex}(\underline{C},\underline{Ab}))^{\circ}$$

(we denote by $\underline{D}^{\circ}$ the dual category of $\underline{D}$), defined by

$$X \longmapsto h^X = \mathrm{Hom}_{\underline{C}}(X, -).$$

In this way $\underline{C}$ can be considered as a subcategory of $(\mathrm{Sex}(\underline{C},\underline{Ab}))^{\circ}$. This embedding is full, i.e.

$$\mathrm{Hom}(X,Y) \xrightarrow{\sim} \mathrm{Hom}(h^X,h^Y),$$

as follows from the Yoneda-lemma (cf. [12], 1.10; cf. EGA 0_{III}.8.1).

Let $\underline{C}$ be an abelian category. Suppose $\underline{C}$ to be artinian (cf. [17], page 355), i.e. every descending chain of subobjects of any object of

$\underline{C}$ is stationnary. Then there exists a canonical equivalence of categories

$$\mathrm{Pro}(\underline{C}) \longrightarrow (\mathrm{Sex}(\underline{C},\underline{Ab}))^{o}$$

(which is "the identity" on $\underline{C} \subset \mathrm{Pro}(\underline{C})$) defined by

$$\{X_i\} \longmapsto \varinjlim_{i} \mathrm{Hom}_{\underline{C}}(X_i, -)$$

(cf. [17], page 356, theorem 1; also compare GP, § 2 and FGA, page 195-06, corollary). It follows that in $\mathrm{Pro}(\underline{C})$ sufficiently many projective objects are available (use [17], page 356, theorem 1, and [18], theorem 1.10.1). Every pro-object is strict (i.e. every object of $\mathrm{Pro}(\underline{C})$ is isomorphic with a projective system in $\underline{C}$ consisting of epimorphisms, cf. FGA, page 195-04, and compare the proof of [17], page 356, theorem 1). If $\{X_i\} \in \mathrm{Pro}(\underline{C})$ is a projective system in $\underline{C}$, and B is an object of $\underline{C}$, then the natural homomorphism

$$\varinjlim_{i} \mathrm{Ext}^n(X_i, B) \xrightarrow{\sim} \mathrm{Ext}^n(X, B) \qquad X = \{X_i\}$$

is an isomorphism for all n ($\mathrm{Ext}^n(-,-)$ calculated in $\mathrm{Pro}(\underline{C})$) (cf. [16], page II-12, corollary 1; cf. GP, page 27, proposition 7). Every artinian object of $\mathrm{Pro}(\underline{C})$ is isomorphic to an object of $\underline{C}$ (cf. the proof of [17], page 356, theorem 1), thus it follows that the full embedding $\underline{C} \subset \mathrm{Pro}(\underline{C})$ is even a "thick" embedding ("épais" in the sense of [18], 1.11; i.e. if $A \longrightarrow B \longrightarrow C$ is an exact sequence in $\mathrm{Pro}(\underline{C})$, with $A, C \in \underline{C}$, then $B \in \underline{C}$).

The general situation just described will be applied to the (artinian) category $\underline{G} = \underline{G}_k$ of commutative k-algebraic group schemes. We write

$$\underline{P} = \mathrm{Pro}(\underline{G}) = (\mathrm{Sex}(\underline{G},\underline{Ab}))^{o}.$$

Objects of $\underline{P}$ will be called pro-algebraic group schemes. Much of GP can be taken over without change; in that case we refer to those results without mentioning that our situation is slightly different.

As the embedding $\underline{G} \subset \underline{P} = \text{Pro}(\underline{G})$ is thick, one can define for $A, B \in \underline{G}$ and $n \geq 0$ a homomorphism

$$\phi^n(A,B): E^n_{\underline{G}}(A,B) \xrightarrow{\sim} E^n_{\underline{P}}(A,B) \cong \text{Ext}^n_{\underline{P}}(A,B)$$

("exact sequences remain exact in $\underline{P}$"). It turns out that the homomorphisms $\phi^n(A,B)$ are isomorphisms (prove that $E^n_{\underline{G}}(-,-)$ are the satellites of $\text{Hom}_{\underline{G}}(-,-)$, and use [13], theorem 3.4.2, or realize that $\underline{G}$ is an artinian category, and use [33], theorem 3.5). Hence in order to calculate (multiple) extension groups of commutative algebraic group schemes, it suffices to determine those extension groups in $\underline{P}$ (and in that category the $\text{Ext}^n(-,-)$ exist). From now on we identify $E^n_{\underline{G}}(A,B)$ and $E^n_{\underline{P}}(A,B)$ for $A, B \in \underline{G}$, and we write $E^n(A,B)$ (and $\text{Ext}(A,B)$ in the case $n = 1$).

REMARK: In GP only projective systems of quasi-algebraic groups were considered (quasi-algebraic groups are objects of the quotient category $\underline{G}/\underline{N}_{loc}$; for the concept of quotient category, cf. [18], 1.11). Therefore it was not possible in that way to determine the extension groups of algebraic groups (cf. GP, page 31, proposition 13).

I.5 The dual of an abelian scheme

For any prescheme Y we write

$$\mathrm{Pic}(Y) = H^1(Y, \underline{O}_Y^*)$$

(where $\underline{O}_Y^*$ denotes the sheaf of multiplicative groups of units). Let S be a prescheme, and X ∈ (Sch/S). We define a contravariant functor

$$F_{X/S}: (\mathrm{Sch}/S)^{\circ} \longrightarrow \underline{\mathrm{Ab}}$$

by

$$F_{X/S}(Y) = \mathrm{Pic}(Y \times_S X), \qquad Y \in (\mathrm{Sch}/S)$$

(for these notions compare FGA, 232 and 236). There is no chance that this functor is representable (in general it is not local, i.e. not a sheaf in the Zariski topology on (Sch/S)). The presheaf (= contravariant functor) $F_{X/S}$ on the category (Sch/S) defines a sheaf in the (fpqc)-topology (we refer to [1] and to SGAD.IV for the definition of a (Grothendieck) topology on a category; (fpqc) stands for "fidèlement plate quasi-compacte", cf. SGAD.IV, page 86). Thus we obtain a functor

$$\underline{P}_{X/S}: (\mathrm{Sch}/S)^{\circ} \longrightarrow \underline{\mathrm{Ab}}$$

(FGA-notation: $\underline{\mathrm{Pic}}_{X/S}$), which is called the <u>Picard functor</u> of X over S. If the Picard scheme of X over S exists (i.e. if $\underline{P}_{X/S}$ is representable), it will be denoted by $\underline{\mathrm{Pic}}(X/S)$. In general it is difficult to describe $\underline{P}_{X/S}$ explicitly. However in special cases it is much easier:

<u>PROPERTY</u> (5.1): Let S be a (locally noetherian) prescheme, and let $f: X \longrightarrow S$ be an <u>abelian scheme</u> over S (compare section 1). The natural map

$$\underline{O}_S \xrightarrow{\sim} f_*(\underline{O}_X)$$

is an isomorphism (cf. FGA, page 232-13, 5.2; cf. EGA, III^2, 7.8.8).

COROLLARY (5.2): Let $X \longrightarrow S$ be an abelian scheme. The Picard functor of X over S is defined by

$$\underline{P}_{X/S}(Y) = \mathrm{Pic}(Y \times_S X) / \mathrm{Pic}(Y).$$

PROOF: Apply FGA, 232-05, 2.4.

Let $X \longrightarrow S$ be a projective abelian scheme. Then the Picard scheme $\underline{\mathrm{Pic}}(X/S)$ exists (cf. FGA, 232, theorem 3.1). In case S = Spec(field), this Picard scheme is reduced, as follows from the fact that the dimension of the vectorspace $H^1(X,\underline{O}_X)$ equals the dimension of X. This fact can be generalized (GROTHENDIECK):

PROPERTY (5.3): Let $X \longrightarrow S$ be an abelian scheme. Suppose that one of the following conditions is true:

a) $X \longrightarrow S$ is projective,

b) S is an artinian scheme (i.e. the spectrum of an artinian ring).

Then

$$X^t \underset{\text{def.}}{=} \underline{\mathrm{Pic}}^{\tau}(X/S) = \underline{\mathrm{Pic}}^{o}(X/S)$$

is an abelian scheme over S, which is projective in case (a). The operation t is a functor (with respect to abelian schemes over S and homomorphisms between them), which commutes with base change.

For the notation G^{τ} and G^{o} we refer to FGA, 236-02. The property $\underline{\mathrm{Pic}}^{\tau}(X/S) = \underline{\mathrm{Pic}}^{o}(X/S)$ holds on fibres as abelian varieties have no torsion (BARSOTTI, cf. [2], theorem 4.3a; cf. [36], theorem 5). For the proof in case (a) we refer to [27], 6.1, corollary 6.8; for the case (a) and (b) compare FGA, 236, sections 3 and 4.

X^t is called the <u>dual abelian scheme</u> of X (perhaps $X^{t/S}$ would be a more correct notation). If $f: X \longrightarrow Y$ is a homomorphism of projective abelian schemes (respectively of abelian schemes over an artinian base), we write

$$f^t : Y^t \longrightarrow X^t$$

for the derived homomorphism. We identify $(X_T)^t$ and $(X^t)_T$ by the natural isomorphism

$$(X_T)^t \xrightarrow{\sim} (X^t)_T$$

obtained by base change $T \longrightarrow S$.

Let X and Y be abelian schemes over S, and let $f: X \longrightarrow Y$ be a homomorphism. We say that f is an <u>isogeny</u> if it is surjective and if Ker(f) is finite over S. In case S = Spec(field) this terminology coincides with the current one. We remark that the kernel of an isogeny is flat over the base (cf. [27] , page 122, lemma 6.12).

If $f: X \longrightarrow Y$ is a surjective homomorphisms of abelian schemes over S such that its kernel has finite fibres, f is an isogeny: Ker(f) is quasi-finite and propre over S (being a closed subscheme of X), hence Ker(f) is finite over S (cf. EGA, III^1.4.4.2). Warning: a group scheme which is quasi-finite over S need not to be finite over S (e.g. take a finite group scheme over S, and delete a point of it such that it remains a group scheme, but such that it does not remain affine).

Chapter II ALGEBRAIC GROUP SCHEMES

We fix the following: k is an algebraically closed field of characteristic $p \neq 0$, $S = \mathrm{Spec}(k)$; $\underline{N} = \underline{N}_S$, the category of commutative finite group schemes over S; $\underline{A} = \underline{A}_k$, the category of finitely generated commutative k-bialgebras; $\underline{G} = \underline{G}_S$, the category of commutative k-algebraic group schemes.

II.6 General facts

GROTHENDIECK has proved the category of commutative algebraic group schemes over a field to be an abelian category (cf. FGA, page 212-17, corollary 7.4, also compare GABRIEL, SGAD.V; no assumptions on k are needed). A homomorphism $\phi : G \longrightarrow H$ of group schemes is epimorphic (in (Sch/S) and in $\underline{G}$) if and only if it is (set-theoretically) surjective and

$$\phi_*(\underline{O}_G) \hookleftarrow \underline{O}_H$$

is injective.

We write $\underline{G}_{red}$ for the full subcategory of $\underline{G}$ consisting of all $G \in \underline{G}$ with $G = G_{red}$ (k is supposed to be algebraically closed, hence perfect)(the objects of $\underline{G}_{red}$ could be called algebraic groups). We denote by $\underline{G}_{con}$, respectively by $\underline{G}_{CR}$, the full subcategory of $\underline{G}$ consisting of connected, respectively connected and reduced, k-algebraic group schemes. For any $G \in \underline{G}$ we denote by $CR(G)$ the largest subobject of G contained in $\underline{G}_{CR}$. Thus we have obtained a functor

$$CR: \underline{G} \longrightarrow \underline{G}_{CR}$$

(which commutes with base extension involving perfect fields).

An abelian variety $A \in \underline{G}$, $A \neq 0$ will be called elementary if $B \subset A$,

$B \in \underline{G}_{CR}$ implies $B = 0$ or $B = A$ (cf LANG, [22], page 29, cf. WEIL, [41]; they used the term "simple abelian variety"; we change terminology as an elementary abelian variety is not a simple object in $\underline{G}$; also we avoid confusion with the term simple = smooth morphism).

DEFINITION (cf. GP, 3.2): The following algebraic group schemes are called elementary:

1) The additive linear group, $\underline{\underline{G}}_a$;
2) the multiplicative linear group, $\underline{\underline{G}}_m$;
3) elementary abelian varieties;
4) the finite group schemes: μ_q, q a prime number, ν_p and α_p

(for definitions, compare pp. I.1 - 2/3 & I.2 - 11/12).

LEMMA (6.1): For any $G \in \underline{G}$ there exist subgroup schemes

$$0 = G_0 \subset G_1 \subset \ldots \subset G_n = G$$

such that G_i/G_{i-1}, $i = 1, 2, \ldots, n$, are elementary group schemes.

PROOF: By the exact sequence

$$0 \longrightarrow CR(G) \longrightarrow G \longrightarrow G/CR(G) \longrightarrow 0$$

it suffices to prove the lemma in the case $G \in \underline{G}_{CR}$ and $G \in \underline{N}$. For the category $\underline{G}_{CR}$ this is known (structure theorems of CHEVALLEY and of BOREL, for references compare GP, 1.3). For the category $\underline{N}$ this is immediate: this category is finite (i.e. noetherian and artinian), thus every $G \in \underline{N}$ can be decomposed into simple objects; the simple objects of $\underline{N}$ are precisely the finite group schemes we call elementary, as follows from the structure theorem of GABRIEL (compare section 2).

QED

LEMMA (6.2): Let $N \subset G$, $N \in \underline{N}_{loc}$, $G \in \underline{G}$. There exists an integer n such that

$$N \subset \underline{\underline{I}}^n(G) \subset G$$

(for the notation $\underline{\underline{I}}^n(-)$, we refer to section 1).

PROOF: Let $\underline{a} = \mathrm{Ker}(\underline{O} \longrightarrow \Gamma(\underline{O}_N))$. As $\underline{O} = \underline{O}_{G,e}$ is a noetherian local ring, there exists an integer n such that $\underline{m}^n \subset \underline{a}$ ($\underline{a}$ is finitely generated and $\bigcap_{i=0}^{\infty} \underline{m}^i = 0$); in that case

$$\underline{m}^{(p^n)} \subset \underline{m}^n \subset \underline{a} \quad \text{thus} \quad N \subset \underline{\underline{I}}^n(G)$$

(cf lemma 1.2).

QED

DEFINITION: An epimorphism $\phi : G \longrightarrow H$ is called a purely inseparable isogeny if $\mathrm{Ker}(\phi) \in \underline{N}_{loc}$.

LEMMA (6.2): Let $\phi : G \longrightarrow H$ be a purely inseparable isogeny. Then there exists a natural number n and a homomorphism $\psi : H \longrightarrow \theta^n G$ such that

$$(G \xrightarrow{\phi} H \xrightarrow{\psi} \theta^n G) = (F_G)^n$$

and

$$(H \xrightarrow{\psi} \theta^n G \xrightarrow{\theta^n \phi} \theta^n H) = (F_H)^n.$$

PROOF: As θ^n is obtained by base change, this functor is exact (cf. FGA, p.212-17, 7.3). Choose n such that for all elements x of the maximal ideal of $\Gamma(\underline{O}_N)$ it holds $x^{p^n} = 0$. There results an exact, commutative diagram

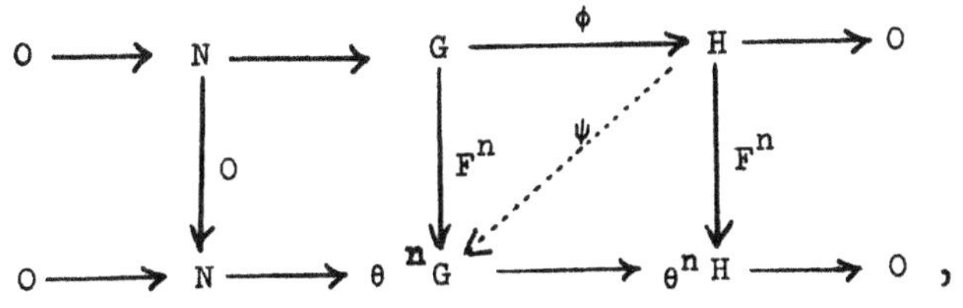

and the lemma follows easily.

We write

$$\underline{P} = \mathrm{Pro}(\underline{G}) = (\,\mathrm{Sex}(\underline{G},\underline{Ab})\,)^{\circ}$$

(cf. section 4). The objects of $\underline{P}$ are called <u>pro-algebraic group schemes</u>. The "groups of dimension zero" form a full subcategory

$$\underline{P}_o = \mathrm{Pro}(\underline{N}),$$

and the pro-algebraic group schemes with topological space equal to one point form a full subcategory

$$\underline{P}_{loc} = \mathrm{Pro}(\underline{N}_{loc}).$$

II.7 The fundamental group

In this section we follow GP very close. We define the functor

$$\pi_0 : \underline{G} \longrightarrow \underline{N} \quad \text{by} \quad \pi_0(G) = G/CR(G)$$

(i.e. $\pi_0(G)$ is the "largest" quotient of G in $\underline{N}$). For any $G \in \underline{G}$ and $N \in \underline{N}$,

$$\mathrm{Hom}(\pi_0 G, N) \xrightarrow{\sim} \mathrm{Hom}(G,N)$$

is an isomorphism. Hence π_0 is left adjoint to the inclusion functor $\underline{N} \subset \underline{G}$, and it follows that π_0 is right exact (cf. [17] , page 341, proposition 11). This functor can be extended to a right exact functor

$$\pi_0 : \underline{P} \longrightarrow \underline{P}_0$$

which commutes with projective limits (cf. GP, 2.7 and 5.1); for $G \in \underline{P}$, $\pi_0(G)$ is the "largest" quotient of G in $\underline{P}_0$ (cf. GP, 5.1). We denote by

$$\pi_i : \underline{P} \longrightarrow \underline{P}_0, \quad i = 0,1,2,\ldots$$

the derived functors of π_0; an exact sequence in $\underline{P}$ yields an exact "homotopy" sequence (cf. GP, 5.3).

The functor $\underline{P} \longrightarrow \underline{P}/\underline{P}_0$ has an adjoint, which we denote by $G \longrightarrow \overline{G}$ (it will be proved that this functor is exact); $\overline{G}$ is called the universal covering of G. It is charaterized by:

$$\overline{G} \in \underline{P}_{CR}, \quad \pi_1(\overline{G}) = 0, \quad \mathrm{Ker}(\overline{G} \longrightarrow G) \in \underline{P}_0, \quad \mathrm{Coker}(\overline{G} \longrightarrow G) \in \underline{P}_0 .$$

The construction of $\overline{G}$ can be carried out with the help of a projective system of isogenies over CR(G) (cf. GP, 6.4).

<u>LEMMA</u> (7.1): Let $N \in \underline{N}_{loc}$ with $\mathrm{Coker}(p.1_N) = 0$. Then $N = 0$.

<u>PROOF</u>: $M = N^D \in \underline{N}_{rl} \bullet \underline{N}_{ll}$ has p-torsion, hence $\mathrm{Ker}(p.1_M) = 0$ implies $M = 0$; as D is exact it follows that $N = 0$. Alternative proof: use Fitting's lemma.

<u>COROLLARY</u> (7.2): Let $N \in \underline{P}_{loc}$ with $\mathrm{Coker}(p.1_N) = 0$. Then $N = 0$.

<u>PROOF</u>: We can choose a strict projective system $\{N_\alpha\}$, $N_\alpha \in \underline{N}_{loc}$, such that $N = \varprojlim N_\alpha$. Let $C_\alpha = \mathrm{Coker}(p.1_{N_\alpha})$. For $\alpha \succ \beta$ we have the following exact, commutative diagram

$$\begin{array}{ccccccc}
N_\alpha & \longrightarrow & N_\alpha & \longrightarrow & C_\alpha & \longrightarrow & 0 \\
\downarrow & & \downarrow & & \downarrow & & \\
N_\beta & \longrightarrow & N_\beta & \longrightarrow & C_\beta & \longrightarrow & 0\,. \\
\downarrow & & \downarrow & & \downarrow & & \\
0 & & 0 & & 0 & &
\end{array}$$

Hence $\{C_\alpha\}$ is a strict projective system. As $\varprojlim C_\alpha = \mathrm{Coker}(p.1_N) = 0$ ($\varprojlim$ is an exact functor in $\underline{P}$), this implies $C_\alpha = 0$ for all α; by (7.1) this implies N_α for all α, hence $N = 0$.

QED

<u>COROLLARY</u> (7.3): For any abelian variety A, $\pi_i(A) = 0$ for $i \geq 2$; also $\pi_i(\underline{\underline{G}}_m) = 0$ for $i \geq 2$ (cf. GP, 6.5, proposition 8, corollary 2 and corollary 3).

<u>PROOF</u>: In case $G = A$, or $G = \underline{\underline{G}}_m$, for every positive integer q the homomorphism $q.1_G$ is an epimorphism. The exact sequence

$$0 \longrightarrow {}_qG \longrightarrow G \xrightarrow{\;q.1_G\;} G \longrightarrow 0$$

proves that multiplication by q is bijective on $\pi_i(G)$ for $i \geq 2$. This implies $\pi_i(G) = 0$ for $i \geq 2$ (for $(q,p) = 1$ this is clear; for $q = p$, use 7.2). This proves the corollary.

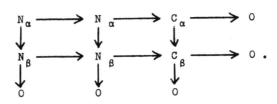

Let $G \in \underline{G}$. The inductive system of finite subgroups of G will be denoted by

$$\Lambda_G = \{N \mid N \subset G, \ N \in \underline{N}\} .$$

PROPOSITION (7.4): Let $X \in \underline{P}_{loc}$ and $H \in \underline{G}$. Then the natural homomorphism

$$\varinjlim_{N \in \Lambda_H} \operatorname{Ext}(X,N) \xrightarrow{\ \sim\ } \operatorname{Ext}(X,H)$$

is an isomorphism.

PROOF: We call this homomorphism α. Let $X = \varprojlim X_i$, $X_i \in \underline{N}_{loc}$. Then

$$\operatorname{Ext}(X,H) = \varinjlim_i \operatorname{Ext}(X_i,H)$$

and

$$\varinjlim_i \varinjlim_{\Lambda_H} \operatorname{Ext}(X_i,N) = \varinjlim_{\Lambda_H} \varinjlim_{i} \operatorname{Ext}(X_i,N) \cong \varinjlim_{\Lambda_H} \operatorname{Ext}(X,N);$$

hence it suffices to consider the case $X \in \underline{N}_{loc}$.

Let $\xi \in \varinjlim_{N \in \Lambda_H} \operatorname{Ext}(X,N)$ such that

$$\begin{array}{lccccccccc}
\xi & 0 & \longrightarrow & N & \longrightarrow & Y & \longrightarrow & X & \longrightarrow & 0 \\
 & & & \big\downarrow & & \big\downarrow & & \| & & \\
\alpha(\xi) = 0 & 0 & \longrightarrow & H & \longrightarrow & G & \underset{\sigma}{\rightleftarrows} & X & \longrightarrow & 0 .
\end{array}$$

In that case there exists a section $\sigma : X \longrightarrow G$. Choose an integer n such that $Y \subset \underline{\underline{I}}^n(G) \subset G$ (cf. 6.2),

$$(X \xrightarrow{\ \sigma'\ } \underline{\underline{I}}^n(G) \hookrightarrow G) = \sigma .$$

Then we obtain a commutative diagram

$$\begin{array}{lccccccccc}
 & 0 & \longrightarrow & N & \longrightarrow & Y & \longrightarrow & X & \longrightarrow & 0 \\
 & & & \big\downarrow & & \big\downarrow & & | & & \\
N \subset M \subset H & 0 & \longrightarrow & M & \longrightarrow & \underline{\underline{I}}^n G & \underset{\sigma'}{\rightleftarrows} & X & \longrightarrow & 0 .
\end{array}$$

Thus $\xi_M = 0$, and we have proved α to be injective.

Suppose given an exact sequence

$$0 \longrightarrow H \longrightarrow G \longrightarrow X \longrightarrow 0, \quad X \in \underline{N}_{loc}.$$

Choose an integer n such that

$$(Z = \underline{\underline{I}}^n(G) \hookrightarrow G \longrightarrow X)$$

is epimorphic ($X = \mathrm{Spec}(A)$, $\underline{O}_{G,e} = B$, $\underline{n}$ the maximal ideal of B, choose n such that $A \cap \underline{n}^{(p^n)} = (0)$). Then there exists a commutative, exact diagram

$$\begin{array}{ccccccccc} 0 & \longrightarrow & N & \longrightarrow & Z & \longrightarrow & X & \longrightarrow & 0 \\ & & \downarrow & & \downarrow & & \| & & \\ 0 & \longrightarrow & H & \longrightarrow & G & \longrightarrow & X & \longrightarrow & 0, \end{array} \qquad N \in \underline{N}_{loc},$$

and we have proved α to be surjective, and the proof is concluded.

<u>COROLLARY</u> (7.5): Let $G \in \underline{P}_0$; G is projective in $\underline{P}_0$ if and only if G is projective in $\underline{P}$. Hence projective envelopes of objects of $\underline{P}_0$ belong to $\underline{P}_0$.

<u>PROOF</u>: $G = G_{red} \oplus G_{loc}$; for the reduced part the arguments of GP, 4.4, can be taken over directly. For $G_{loc} = X \in \underline{P}_{loc}$, and for all $H \in \underline{G}$,

$$\varinjlim \mathrm{Ext}(X,N) \xrightarrow{\sim} \mathrm{Ext}(X,H)$$

is an isomorphism (cf. 7.4). If X is projective in $\underline{P}_0$, this implies $\mathrm{Ext}(X,H) = 0$ for all $H \in \underline{G}$, hence X is projective in $\underline{P}$ (cf. GP, 3.1, proposition 2), and the corollary is proved.

<u>REMARK</u>: Another way of stating this result: $\underline{P}_0$ is a colocalizing subcategory of $\underline{P}$ (cf. [17] , page 377, corollary of proposition 8).

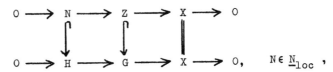

Thus it follows that for every $G \in \underline{G}$ the sequence

$$0 \longrightarrow \pi_1(G) \longrightarrow \overline{G} \longrightarrow G \longrightarrow \pi_0(G) \longrightarrow 0$$

is exact (cf GP, page 40, proposition 3).

Let q be a prime number (q = p is not excluded). We denote by $\mathbb{Z}_q$ the zero-dimensional proalgebraic group scheme:

$$\mathbb{Z}_r = \varprojlim_i (\mu_{r^i}) \ , \quad r \text{ a prime different from } p,$$

$$\mathbb{Z}_p = \varprojlim_i (\nu_{p^i})$$

(cf GP, 4.3). These groups are projective objects (as well in $\underline{P}_0$ as in $\underline{P}$);

$$0 \longrightarrow \mathbb{Z}_r \xrightarrow{\times r} \mathbb{Z}_r \longrightarrow \mu_r \longrightarrow 0$$

is a projective resolution for μ_r, $(r,p) = 1$, and

$$0 \longrightarrow \mathbb{Z}_p \xrightarrow{\times p} \mathbb{Z}_p \longrightarrow \nu_p \longrightarrow 0$$

is a projective resolution for ν_p.

II.8 Multiplicative group schemes (linear and infinitesimal)

For every positive integer n we define K_n by the exact sequence

$$0 \longrightarrow K_n \longrightarrow \underline{\underline{G}}_m \xrightarrow{F^n} \underline{\underline{G}}_m \longrightarrow 0 .$$

Clearly

$$K_n \in \underline{N}_{1r} , \quad K_n = \underline{\underline{I}}^n(\underline{\underline{G}}_m).$$

LEMMA (8.1): For all $n \geq 1$, $K_n^D = \nu_{p^n}$.

PROOF: As $K_1 = \mu_p$, we have $K_1^D = \nu_p$. Consider the exact commutative diagram

$$\begin{array}{ccccccccc}
 & & 0 & & 0 & & & & \\
 & & \downarrow & & \downarrow & & & & \\
 & & K_1 & \xrightarrow{\sim} & K_1 & & & & \\
 & & \downarrow & & \downarrow & & & & \\
0 & \longrightarrow & K_{n+1} & \longrightarrow & \underline{\underline{G}}_m & \xrightarrow{F^{n+1}} & \underline{\underline{G}}_m & \longrightarrow & 0 \\
 & & \downarrow{\scriptstyle \phi_{n+1}} & & \downarrow{\scriptstyle F} & & \| & & \\
0 & \longrightarrow & K_n & \longrightarrow & \underline{\underline{G}}_m & \xrightarrow{F^{n}} & \underline{\underline{G}}_m & \longrightarrow & 0 \\
 & & \downarrow & & \downarrow & & & & \\
 & & 0 & & 0 & & & &
\end{array}$$

It follows that K_{n+1} is an extension over K_n with kernel K_1. This extension is not split, as multiplication by p^n is zero on K_1 and on K_n, but it is not zero on K_{n+1}. Assume (induction) that $(K_n^D)_k \cong \mathbb{Z}/p^n\mathbb{Z}$. An extension over $(K_1^D)_k \cong \mathbb{Z}/p\,\mathbb{Z}$ with kernel $\mathbb{Z}/p^n\,\mathbb{Z}$ which is not split, is isomorphic with $\mathbb{Z}/p^{n+1}\,\mathbb{Z}$. Hence the lemma is proved by induction.

Alternative proof: direct calculation, or application of the Cartier-Shatz formula (cf. section 16).

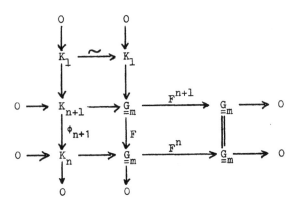

We denote by $K_\infty \in \mathrm{Pro}(\underline{N})$ the projective limit $K_\infty = \varprojlim K_n$, where the morphisms

$$\phi_{n+1} : K_{n+1} \longrightarrow K_n,$$

$n \geq 1$, defined above are used.

COROLLARY (8.2): K_∞ is the projective envelope of $K_1 = \mu_p$, and

$$0 \longrightarrow K_\infty \xrightarrow{\times p} K_\infty \longrightarrow \mu_p \longrightarrow 0$$

is a projective resolution of μ_p.

PROOF: This follows, using (8.1), from the fact that $\varinjlim \mathbb{Z}/p^n\mathbb{Z}$ is the injective hull of $\mathbb{Z}/p\mathbb{Z}$ in $\underline{Ab}$ (hence K_∞ is projective in $\underline{P}_0$, and use 7.5).

QED

PROPOSITION (8.3): $\pi_1(\underline{\underline{G}}_m) = K_\infty \oplus \prod_{q \neq p} \mathbb{Z}_q$, $\overline{\underline{\underline{G}}}_m \longrightarrow \underline{\underline{G}}_m$ is the projective envelope of $\underline{\underline{G}}_m$, and

$$0 \longrightarrow \pi_1(\underline{\underline{G}}_m) \longrightarrow \overline{\underline{\underline{G}}}_m \longrightarrow \underline{\underline{G}}_m \longrightarrow 0$$

is a projective resolution for $\underline{\underline{G}}_m$; hence $dp(\underline{\underline{G}}_m) = 1$.

PROOF: The structure of $(\pi_1(\underline{\underline{G}}_m))_{red}$ follows directly from GP, 6.5, corollary 2. The structure of $(\pi_1(\underline{\underline{G}}_m))_{loc}$ follows from the construction of $\overline{\underline{\underline{G}}}_m$ (and from the definition of K_∞). Further GP, 7.4, corollary 2 and GP, 7.5, proposition 5 can be taken over.

QED

Let $G \in \underline{G}$. We can write $\underline{\underline{I}}^1(G) = \mu_p^\sigma \oplus N$, with $N \in \underline{N}_{11}$, as $(((\underline{\underline{I}}^1 G)^D)_{rl})_k$ has only points with order dividing p.

DEFINITION: We write $\sigma = \sigma(G)$ if $(\underline{\underline{I}}^1 G)_{1r} = \mu_p^\sigma$.

EXAMPLES: $\sigma(\underline{\underline{G}}_m) = 1 = \sigma(\mu_p)$; $\sigma(\underline{\underline{G}}_a) = 0 = \sigma(\alpha_p) = \sigma(H_q)$; $0 \leq \sigma(A) \leq \dim A$, if A is a reduced algebraic group scheme.

NOTATION: $\varinjlim K_n = K^{\infty} \in \mathrm{Ind}(\underline{N})$; for $G \in \underline{G}$ we write

$$\pi^1(G) = \varinjlim \underline{I}^n(G) \in \mathrm{Ind}(\underline{N})$$

(hence $K^{\infty} = \pi^1(\underline{\underline{G}}_m)$; notation of [23] : $\pi^1(G) = \hat{G}$).

LEMMA (8.4): Let $G = G_{red} \in \underline{G}$. Then

$$\varinjlim_n (\underline{\underline{I}}^n G)_{1r} = (K^{\infty})^{\sigma(G)}.$$

PROOF: As $\underline{N}_{1r}$ is equivalent with the category of finite p-groups, $(\pi^1 G)_{1r}$ can be written as a direct sum of copies of K and of objects K_n. Suppose $(\pi^1 G)_{1r} = N \oplus K_n$; in that case K_n is a direct summand of G, in contradiction with $G = G_{red}$. Hence $(\pi^1 G)_{1r}$ is a direct sum of copies of K ; clearly there are $\sigma(G)$ copies, and the lemma is proved.

THEOREM (8.5): For every reduced $G \in \underline{G}$, $\mathrm{Ext}(\mu_p, G) = 0$.
[GROTHENDIECK communicated to me, that he proved this theorem in a more general form.]

PROOF:

$$\mathrm{Ext}(\mu_p, G) \cong \varinjlim_{\Lambda_G} \mathrm{Ext}(\mu_p, N)$$

(cf. 7.4), and

$$\varinjlim \mathrm{Ext}(\mu_p, N) \cong \mathrm{Ext}_{\mathrm{Ind}(\underline{N})}(\mu_p, (K^{\infty})^{\sigma(G)})$$

(cf. [16] , page II-12, corollary 1). As K^{∞} is an injective object in $\mathrm{Ind}(\underline{N})$, the theorem is proved.

COROLLARY (8.6): Let $f: G \longrightarrow G'$ be an isogeny of reduced algebraic group schemes. Then $\sigma(G) = \sigma(G')$.

PROOF: It suffices to prove the case that Ker(f) is a simple object of $\underline{N}$. Let $N = \mathrm{Ker}(f)$; consider the exact sequence

$$0 \longrightarrow \mathrm{Hom}(\mu_p, N) \longrightarrow \mathrm{Hom}(\mu_p, G) \xrightarrow{f_*} \mathrm{Hom}(\mu_p, G') \longrightarrow \mathrm{Ext}(\mu_p, N) \longrightarrow$$

In the cases $N = \alpha_p$, $N = \mu_q$ with $(q,p) = 1$, $N = \nu_p$, the homomorphism f_* is an isomorphism. If $N = \mu_p$,

$$|\mathrm{Hom}(\mu_p, \mu_p)| = p = |\mathrm{Ext}(\mu_p, \mu_p)|$$

(notation: $|..|$ is the number of elements of the group ..); hence in all cases

$$p^{\sigma(G)} = |\mathrm{Hom}(\mu_p, G)| = |\mathrm{Hom}(\mu_p, G')| = p^{\sigma(G')} \quad ,$$

and the corollary is proved.

COROLLARY (8.7): For all $G \in \underline{P}$ and all $n \geq 2$:

$$E^n(G, \mu_p) = 0.$$

PROOF: It suffices to prove the case $n = 2$. As $E^2(G/CR(G), \mu_p) = 0$, it suffices to consider only $G \in \underline{G}_{CR}$. Let $\xi \in E^2(G, \mu_p)$ be given by an exact sequence

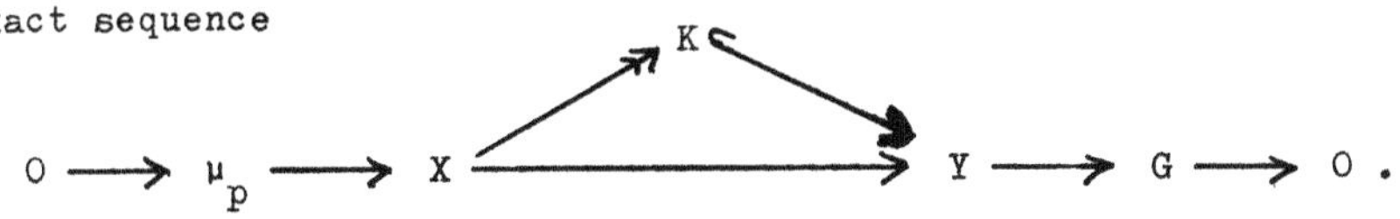

As $\mathrm{Ext}(K_{red}, \mu_p) = 0$ (cf. 8.5) we can construct a commutative, exact diagram with $L = K/K_{red}$:

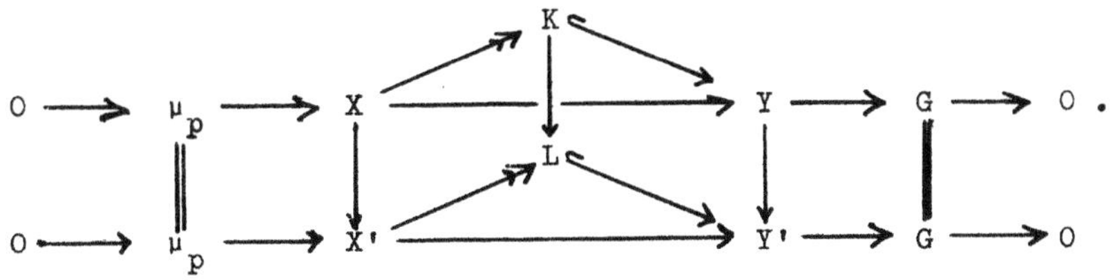

As $\mathrm{Ext}(L_{red}, \mu_p) = 0 = \mathrm{Ext}(L_{11}, \mu_p)$ and $\mathrm{Ext}(G, L_{1r})$ we have proved $\xi = 0$.

QED

PROPOSITION (8.8): For any $G \in \underline{G}_{CR}$, $(\pi_1(G))_{1r} = (K_\infty)^{\sigma(G)}$.

PROOF: The exact sequence

$$0 \longrightarrow \pi_1(G) \longrightarrow \overline{G} \longrightarrow G \longrightarrow 0$$

yields an exact sequence

$$0 \longrightarrow \mathrm{Hom}(\pi_1(G), \mu_p) \longrightarrow \mathrm{Ext}(G, \mu_p) \longrightarrow$$
$$\longrightarrow \mathrm{Ext}(\overline{G}, \mu_p) \longrightarrow \mathrm{Ext}(\pi_1(G), \mu_p) \longrightarrow E^2(G, \mu_p) .$$

As $\mathrm{Ext}(\overline{G}, \mu_p) = 0$ and $E^2(G, \mu_p) = 0$, it follows

$$\mathrm{Ext}(\pi_1(G), \mu_p) = 0 ,$$

hence $(\pi_1(G))_{1r}$ is a projective object (it is projective in $\underline{P}_0$ as μ_p is the only simple object of $\underline{N}_{1r}$, and use 7.5). The only projective objects in $\underline{P}_{1r}$ are direct products of copies of K (cf. [15] , pp. 17 - 11/12). As

$$\mathrm{Hom}(\pi_1(G), \mu_p) \xrightarrow{\sim} \mathrm{Ext}(G, \mu_p)$$

is an isomorphism, and

$$|\mathrm{Ext}(G, \mu_p)| = p^{\sigma(G)}$$

the proposition is proved.

II.9 Witt groups

In this section we recall some results, essentially due to SERRE (cf. GA, VII.8-10).

The underlying scheme of the Witt group scheme $W_n \in \underline{G}$ is given by $W_n = \mathrm{Spec}(k[X_1,\dots,X_n])$, while the group structure is defined by Witt polynomials (for references, cf. GA, VII.8; also compare [37] and [16]) ; $W_1 = \underline{\underline{G}}_a$. As W_n can be defined over the prime field $\theta(W_n) = W_n$, hence the Frobenius homomorphisms $F : W_n \longrightarrow W_n$ exist; further the homomorphisms $V : W_n \longrightarrow W_{n+1}$ ("Verschiebung") and $R : W_{n+1} \longrightarrow W_n$ (restriction) are well-known; these homomorphisms commute. We denote the ring $\mathrm{Hom}(W_n,W_n)$ by A_n: the (non-commutative) ring A_1 consists of "polynomials" over k in the letter F, while multiplication is given by

$$F\alpha = \alpha^p F \qquad \alpha \in k.$$

The extension

$$0 \longrightarrow W_m \xrightarrow{V^n} W_{n+m} \xrightarrow{R^m} W_n \longrightarrow 0$$

defines an element $\alpha_n^m \in \mathrm{Ext}(W_n,W_m)$. For reference we state:

LEMMA (cf. GA, VII.9)(9.1):

a) The homomorphism

$$\delta_1 : \mathrm{Hom}(\underline{\underline{G}}_a,\underline{\underline{G}}_a) \xrightarrow{\sim} \mathrm{Ext}(\underline{\underline{G}}_a,\underline{\underline{G}}_a)$$

defined by $\delta_1(\psi) = \psi^*(\alpha_1^1) = \alpha_1^1 \cdot \psi$ is an isomorphism;

b)

$$\delta_2 : \mathrm{Hom}(\underline{\underline{G}}_a, \underline{\underline{G}}_a) \xrightarrow{\sim} \mathrm{Ext}(\underline{\underline{G}}_a,\underline{\underline{G}}_a)$$

defined by $\delta_2(\phi) = \phi_*(\alpha_1^1) = \phi \cdot \alpha_1^1$ is an isomorphism;

c) $V^n : W_m \longrightarrow W_{m+n}$ induces an isomorphism

$$(V^n)^* : \mathrm{Ext}(W_{m+n},\underline{\underline{G}}_a) \xrightarrow{\sim} \mathrm{Ext}(W_m,\underline{\underline{G}}_a),$$

d) and the zero map

$$[(V^n)_* : \mathrm{Ext}(\underline{\underline{G}}_a, W_m) \longrightarrow \mathrm{Ext}(\underline{\underline{G}}_a, W_{m+n})] = 0$$

for every positive m and n;

e) $R^n : W_{m+n} \longrightarrow W_m$ induces an isomorphism

$$(R^n)_* : \mathrm{Ext}(\underline{\underline{G}}_a, W_{m+n}) \xrightarrow{\sim} \mathrm{Ext}(\underline{\underline{G}}_a, W_m),$$

f) and the zero map

$$[(R^n)^* : \mathrm{Ext}(W_m, \underline{\underline{G}}_a) \longrightarrow \mathrm{Ext}(W_{m+n}, \underline{\underline{G}}_a)] = 0$$

for every positive m and n.

<u>COROLLARY</u> (9.2): $F : \underline{\underline{G}}_a \longrightarrow \underline{\underline{G}}_a$ induces an injection

$$F_* : \mathrm{Ext}(W_n, \underline{\underline{G}}_a) \hookrightarrow \mathrm{Ext}(W_n, \underline{\underline{G}}_a).$$

<u>PROOF</u>: F is not a zero divisor in the ring A_1, and use (9.1a) and (9.1c).

<u>COROLLARY</u> (9.3):

$$F^* : \mathrm{Ext}(W_n, \underline{\underline{G}}_a) \hookrightarrow \mathrm{Ext}(W_n, \underline{\underline{G}}_a).$$

<u>PROOF</u>: The case $n = 1$ follows by (9.1a); using (9.1c) the general case then follows.

The group schemes $W_n \in \underline{G}$ form a projective system with respect to the morphisms $R^n : W_{m+n} \longrightarrow W_m$, $m \geq 1$, $n \geq 1$. Its limit we denote by

$$W = \varprojlim_n W_n, \qquad W \in \underline{P}.$$

<u>COROLLARY</u> (9.4):

$$\mathrm{Ext}(W, \underline{\underline{G}}_a) = 0.$$

[cf. GP, 8.5, theorem 1; the proof given here could have been used in GP.]

PROOF:

$$\mathrm{Ext}(W,\underline{\underline{G}}_a) \cong \varinjlim \mathrm{Ext}(W_n,\underline{\underline{G}}_a),$$

and use (9.1f).

QED

COROLLARY (9.5): a) For every $m > 0$ and $n \geq 0$ the homomorphism $R^n : W_{m+n} \longrightarrow W_m$ induces an isomorphism

$$(R^n)^* : \mathrm{Hom}(W_m,\underline{\underline{G}}_a) \xrightarrow{\sim} \mathrm{Hom}(W_{m+n},\underline{\underline{G}}_a)$$

(or: every $W_{m+n} \longrightarrow \underline{\underline{G}}_a$ is factored by R^n);

b) for every $m > 0$ and $n \geq 0$ the homomorphism $V^n : W_m \longrightarrow W_{m+n}$ induces an isomorphism

$$(V^n)_* : \mathrm{Hom}(\underline{\underline{G}}_a,W_m) \xrightarrow{\sim} \mathrm{Hom}(\underline{\underline{G}}_a,W_{m+n})$$

(or: every $\underline{\underline{G}}_a \longrightarrow W_{m+n}$ factors through $W_m \subset W_{m+n}$).

PROOF: From (9.1a) it follows that

$$R^* : \mathrm{Hom}(\underline{\underline{G}}_a,\underline{\underline{G}}_a) = A_1 \xrightarrow{\sim} \mathrm{Hom}(W_2,\underline{\underline{G}}_a)$$

is an isomorphism. Suppose that

$$(R^{n-1})^* : A_1 \xrightarrow{\sim} \mathrm{Hom}(W_n,\underline{\underline{G}}_a)$$

is an isomorphism for some $n \geq 2$. The commutative diagram

$$\begin{array}{ccccccccc}
0 & \longrightarrow & W_n & \xrightarrow{V} & W_{n+1} & \xrightarrow{R^n} & \underline{\underline{G}}_a & \longrightarrow & 0 \\
 & & \downarrow{\scriptstyle R^{n-1}} & & \downarrow{\scriptstyle R^{n-1}} & & \| & & \\
0 & \longrightarrow & \underline{\underline{G}}_a & \xrightarrow{V} & W_2 & \xrightarrow{R} & \underline{\underline{G}}_a & \longrightarrow & 0
\end{array}$$

yields a commutative diagram

$$\begin{array}{ccccccccc}
0 \longrightarrow & A_1 & \xrightarrow{(R^n)^*} & \mathrm{Hom}(W_{n+1},\underline{\underline{G}}_a) & \longrightarrow & \mathrm{Hom}(W_n,\underline{\underline{G}}_a) & \xrightarrow{\delta} & \mathrm{Ext}(\underline{\underline{G}}_a,\underline{\underline{G}}_a) \\
 & \Vert & & \uparrow & & \cong\uparrow (R^{n-1})^* & & \Vert \\
0 \longrightarrow & A_1 & \longrightarrow & \mathrm{Hom}(W_2\,,\underline{\underline{G}}_a) & \longrightarrow & A_1 & \xrightarrow[\sim]{\delta_1} & \mathrm{Ext}(\underline{\underline{G}}_a,\underline{\underline{G}}_a)
\end{array}$$

with exact rows. Induction hypothesis and (9.1a) prove δ to be an isomorphism, hence

$$(R^n)^* : A_1 \xrightarrow{\sim} \mathrm{Hom}(W_{n+1},\underline{\underline{G}}_a)$$

is an isomorphism. The first part of the corollary now follows from the commutative diagram

$$\begin{array}{ccc}
 & A_1 & \\
\swarrow \sim\ (R^{m-1})^* & & (R^{m+n-1})^* \sim \searrow \\
(R^n)^* : \mathrm{Hom}(W_m,\underline{\underline{G}}_a) & \longrightarrow & \mathrm{Hom}(W_{m+n},\underline{\underline{G}}_a).
\end{array}$$

The second part follows analogously, using the diagram

$$\begin{array}{ccccccc}
0 \longrightarrow & \underline{\underline{G}}_a & \xrightarrow{V^n} & W_{n+1} & \xrightarrow{R} & W_n & \longrightarrow 0 \\
 & \Vert & & \uparrow V^{n-1} & & \uparrow V^{n-1} & \\
0 \longrightarrow & \underline{\underline{G}}_a & \xrightarrow{V} & W_2 & \xrightarrow{R} & \underline{\underline{G}}_a & \longrightarrow 0\,.
\end{array}$$

QED

<u>COROLLARY</u> (9.6): $\mathrm{Im}(F^* : \mathrm{Hom}(W_n,\underline{\underline{G}}_a) \longrightarrow \mathrm{Hom}(W_n,\underline{\underline{G}}_a)) =$

$= \mathrm{Im}(F_* : \mathrm{Hom}(W_n,\underline{\underline{G}}_a) \longrightarrow \mathrm{Hom}(W_n,\underline{\underline{G}}_a)).$

<u>PROOF</u>: As the ground field is perfect $A_1 \cdot F = F \cdot A_1$, and the corollary follows using 9.5a (or using 9.5b).

<u>COROLLARY</u> (9.7) (cf. GA, VII.9, lemma 5):

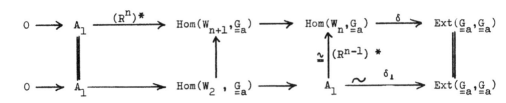

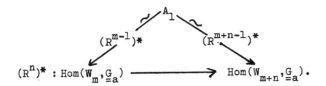

a) For every $n \geq 1$ and every $\phi \in A_n$ there exists a $\psi \in A_{n+1}$ such that $\phi R = R \psi$.

b) For every $n \geq 1$ and every $\rho \in A_n$ there exists a $\tau \in A_{n+1}$ such that $V\rho = \tau V$.

<u>PROOF</u>: a) The exact sequence

$$0 \longrightarrow \underline{\underline{G}}_a \xrightarrow{\;V^n\;} W_{n+1} \xrightarrow{\;R\;} W_n \longrightarrow 0$$

induces a commutative exact diagram

$$\begin{array}{ccccc}
 & & \mathrm{Ext}(W_n, \underline{\underline{G}}_a) & \xrightarrow{\;R^* = 0\;} & \mathrm{Ext}(W_{n+1}, \underline{\underline{G}}_a) \\
 & & \uparrow{\scriptstyle \delta} & & \uparrow{\scriptstyle \delta} \\
0 & \longrightarrow & \mathrm{Hom}(W_n, W_n) = A_n & \xrightarrow{\;R^*\;} & \mathrm{Hom}(W_{n+1}, W_n) \\
 & & & & \uparrow{\scriptstyle R_*} \\
 & & & & A_{n+1} = \mathrm{Hom}(W_{n+1}, W_{n+1}).
\end{array}$$

Using (9.1f) the first part follows by diagram chasing.

b) The exact sequence

$$0 \longrightarrow W_n \xrightarrow{\;V\;} W_{n+1} \xrightarrow{\;R^n\;} \underline{\underline{G}}_a \longrightarrow 0$$

yields

$$\begin{array}{ccccc}
 & & \mathrm{Ext}(\underline{\underline{G}}_a, W_n) & \xrightarrow{\;V_* = 0\;} & \mathrm{Ext}(\underline{\underline{G}}_a, W_{n+1}) \\
 & & \uparrow{\scriptstyle \delta} & & \uparrow{\scriptstyle \delta} \\
0 & \longrightarrow & A_n & \xrightarrow{\;V_*\;} & \mathrm{Hom}(W_n, W_{n+1}) \\
 & & & & \uparrow{\scriptstyle V^*} \\
 & & & & A_{n+1}
\end{array}$$

and analogous arguments apply; thus the proof is concluded.

<u>LEMMA</u> (9.8): $\mathrm{Ext}(\nu_p, \underline{\underline{G}}_a) \cong k$ (i.e. the left A_1-module $A_1/(F) \cong k^+$), and for every $\alpha \in k$, $\alpha \neq 0$, $\alpha - F = \phi_\alpha \in A_1$, the exact sequence

$$0 \longrightarrow \nu_p \xrightarrow{i_\alpha} \underline{\underline{G}}_a \xrightarrow{\phi_\alpha} \underline{\underline{G}}_a \longrightarrow 0$$

induces a surjective homomorphism

$$\mathrm{Ext}(\underline{\underline{G}}_a, \underline{\underline{G}}_a) \xrightarrow{(i_\alpha)^*} \mathrm{Ext}(\nu_p, \underline{\underline{G}}_a).$$

PROOF: We fix a generator $a \in (\nu_p)_k$. If X is an extension over ν_p with kernel $\underline{\underline{G}}_a$, the group scheme X is the disjoint union $X = X_1 \cup \ldots \cup X_p$ of copies of X, where X_i maps onto a^i, $1 \leq i \leq p$. Multiplication by p maps X_1 onto a point $\beta \in (X_p)_k \cong k$ (X_1 is a homogeneous space over $X_p = \underline{\underline{G}}_a$). Thus we have associated with every element of $\mathrm{Ext}(\nu_p, \underline{\underline{G}}_a)$ an element of k. Direct verification proves this map to be an isomorphism (or: both are canonically isomorphic with $A_1/(F)$).

Easy calculation proves

$$\mathrm{Coker}((i_\alpha)_* : A_1 \longrightarrow A_1) \cong k,$$

and the second part of the lemma follows from the first using (9.1b).

QED

REMARK: The lemma can also be deduced from (7.4).

LEMMA (9.9): Let $G \in \underline{G}$ and let $f: G \longrightarrow \underline{\underline{G}}_a$ be an epimorphism. Then there exists a natural number n and a commutative diagram

$$\begin{array}{ccc} G & \xrightarrow{f} & \underline{\underline{G}}_a \\ h \uparrow & \nearrow g & \\ W_n & & \end{array}$$

such that g is an epimorphism.

PROOF: Let L be the maximal unipotent subgroup of G_{red}. Then the composition

$$(L \hookrightarrow G \xrightarrow{f} \underline{\underline{G}}_a)$$

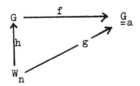

is an epimorphism, as $\mathrm{Hom}(\underline{\underline{G}}_m, \underline{\underline{G}}_a) = 0$ and $\mathrm{Hom}(A, \underline{\underline{G}}_a) = 0$ for every abelian variety A. There exists an epimorphism

$$W_{n_1} \oplus \ldots \oplus W_{n_s} \longrightarrow L$$

(cf. GA, VII.13, theorem 3), and the lemma follows easily.

REMARK: For a description of the rings A_n, compare section 15.

II.10 Infinitesimal Witt groups

Let m and n be positive integers. We define

$$L_{n,m} = \mathrm{Ker}(W_n \xrightarrow{F^m} W_n) = \underline{\underline{I}}^m(W_n)\ , \quad L_{1,1} = \alpha_p .$$

If $m' \geq m$, $n' \geq n$, we define

$$\rho = \rho(m',n';m,n) : L_{n',m'} \longrightarrow L_{n,m}$$

by the exact commutative diagram ($s = n'-n$, $t = m'-m$)

$$\begin{array}{ccccccccc} 0 & \longrightarrow & L_{n,m} & \longrightarrow & W_n & \xrightarrow{F^m} & W_n & \longrightarrow & 0 \\ & & \uparrow{\scriptstyle\rho} & & \uparrow{\scriptstyle R^sF^t} & & \uparrow{\scriptstyle R^s} & & \\ 0 & \longrightarrow & L_{n',m'} & \longrightarrow & W_{n'} & \xrightarrow{F^{m'}} & W_{n'} & \longrightarrow & 0\ ; \end{array}$$

clearly ρ is an epimorphism. We write $F = \rho(m+1,n;m,n)$ and $R = \rho(m,n+1;m,n)$. We have thus defined a projective system, its limit (in $\underline{P}$ or in $\underline{P}_0$) we denote by

$$L_{\infty,\infty} = \varprojlim_{n,m} L_{n,m} .$$

The homomorphism $V : L_{n,m} \longrightarrow L_{n+1,m}$ we define by the exact commutative diagram

$$\begin{array}{ccccccccc} 0 & \longrightarrow & L_{n,m} & \longrightarrow & W_n & \xrightarrow{F^m} & W_n & \longrightarrow & 0 \\ & & \downarrow{\scriptstyle V} & & \downarrow{\scriptstyle V} & & \downarrow{\scriptstyle V} & & \\ 0 & \longrightarrow & L_{n+1,m} & \longrightarrow & W_{n+1} & \xrightarrow{F^m} & W_{n+1} & \longrightarrow & 0. \end{array}$$

Clearly V is a monomorphism for every n and m. The natural inclusion

$$\underline{\underline{I}}^m(W_n) = L_{n,m} \subset L_{n,m+1} = \underline{\underline{I}}^{m+1}(W_n)$$

will be denoted by $i : L_{n,m} \longrightarrow L_{n,m+1}$.

<u>LEMMA</u> (10.1): A homomorphism $f\colon L_{n,1} \longrightarrow L_{1,2}$ is not epimorphic.

<u>PROOF</u>: We write $E_{m,n} = (\underline{O}_{L_{m,n}})$, and by $\underline{m}_{m,n}$ we denote the maximal ideal of $E_{m,n}$. For every $x \in \underline{m}_{n,1}$, $x^p = 0$, hence $E_{1,2} \longrightarrow E_{n,1}$ factors through $E_{1,2} \longrightarrow E_{1,1}$. Thus $E_{1,2} \longrightarrow E_{n,1}$ is not monomorphic, and the lemma is proved.

<u>LEMMA</u> (10.2): For every $n \geq 1$ and every $m \geq 1$, the extensions

$$0 \longrightarrow \alpha_p \xrightarrow{\ i\ } L_{1,m+1} \xrightarrow{\ F\ } L_{1,m} \longrightarrow 0$$

and

$$0 \longrightarrow \alpha_p \xrightarrow{\ V^n\ } L_{n+1,1} \xrightarrow{\ R\ } L_{n,1} \longrightarrow 0$$

are not split.

<u>PROOF</u>: For $x \in \underline{m}_{1,m}$, it holds $x^{p^m} = 0$, and there exists an element $y \in \underline{m}_{1,m+1}$ such that $y^{p^m} \neq 0$; thus the first sequence is not split.

Consider the exact, commutative diagram

$$\begin{array}{lccccccccc}
 & & & 0 & & 0 & & 0 & & \\
 & & & \downarrow & & \downarrow & & \downarrow & & \\
\tau & 0 & \longrightarrow & \alpha_p & \longrightarrow & L_{n+1,1} & \longrightarrow & L_{n,1} & \longrightarrow & 0 \\
 & & & \downarrow j_1 & & \downarrow & & \downarrow j_n & & \\
\alpha^1_n & 0 & \longrightarrow & \underline{\underline{G}}_a & \longrightarrow & W_{n+1} & \longrightarrow & W_n & \longrightarrow & 0. \\
 & & & & & & & \downarrow F & & \\
 & & & & & & & W_n & & \\
 & & & & & & & \downarrow & & \\
 & & & & & & & 0 & &
\end{array}$$

Using (9.1c) it is easily seen that $\alpha^1_n \notin \operatorname{Im}(F^*\colon \operatorname{Ext}(W_n, \underline{\underline{G}}_a) \longrightarrow \operatorname{Ext}(W_n, \underline{\underline{G}}_a))$, hence $0 \neq (j_n)^*(\alpha^1_n) = (j_1)_*(\tau)$; thus $\tau \neq 0$, and the second part of the lemma is also proved.

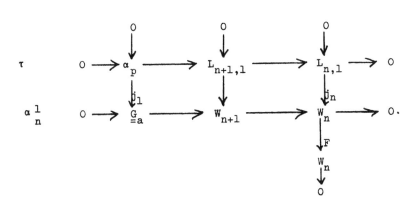

REMARK: The second part of this lemma can be deduced from the first part in the following way: $(L_{n,m})^D \cong L_{m,n}$ and

$$(L_{n+1,m} \xrightarrow{R} L_{n,m})^D = (L_{m,n+1} \xrightarrow{i} L_{m,n}),$$

$$(L_{n,m+1} \xrightarrow{F} L_{n,m})^D = (L_{m+1,n} \xrightarrow{V} L_{m,n})$$

correspond in this way under these isomorphisms; dualizing

$$0 \longrightarrow \alpha_p \xrightarrow{V^n} L_{n+1,1} \xrightarrow{R} L_{n,1} \longrightarrow 0$$

we obtain:

$$0 \longleftarrow \alpha_p \xleftarrow{F^n} L_{1,n+1} \xleftarrow{i} L_{1,n} \longleftarrow 0$$

and this sequence is clearly not split.

PROPOSITION (10.3): The extensions $L_{n,1} \xrightarrow{\rho} \alpha_p$, $L_{1,m} \xrightarrow{\rho} \alpha_p$ are essential for $n \geq 1$, $m \geq 1$.

PROOF: Suppose that $\rho : L_{n,1} \longrightarrow \alpha_p$ is essential for some $n \geq 1$ (for $n = 1$ this is trivial). Let $X \subset L_{n+1,1}$ such that

$$(X \longrightarrow L_{n+1,1} \xrightarrow{\rho} \alpha_p)$$

is an epimorphism. Then $(X \longrightarrow L_{n+1,1} \longrightarrow L_{n,1})$ is an epimorphism by the fact that $L_{n,1} \longrightarrow \alpha_p$ is essential. Thus we obtain a commutative, exact diagram

$$\begin{array}{ccccccccc}
0 & \longrightarrow & \alpha_p & \longrightarrow & L_{n+1,1} & \longrightarrow & L_{n,1} & \longrightarrow & 0 \\
 & & \uparrow \phi & & \cup & & \| & & \\
0 & \longrightarrow & K & \longrightarrow & X & \longrightarrow & L_{n,1} & \longrightarrow & 0.
\end{array}$$

By the five lemma, ϕ is monomorphic. As α_p is a simple object, this implies $K = 0$ or $K = \alpha_p$. In the first case the extension

$L_{n+1,1}/\alpha_p \cong L_{n,1}$ is split, which is a contradiction by (10.2). Hence $K = \alpha_p$, thus $X = L_{n+1,1}$, and we proceed by induction.

The second statement is proved analogously (it can also be derived from the first with the help of the preceding remark).

<u>PROPOSITION</u> (10.4): For every m and n the extension

$$L_{n,m} \longrightarrow \alpha_p$$

is essential.

<u>PROOF</u>: We proceed by induction on m: suppose that for some $m \geq 1$ and for all $n \geq 1$ the extension $L_{n,m} \longrightarrow \alpha_p$ is essential. For some $n \geq 1$ we choose $X \subset L_{n,m+1}$ such that $(X \longrightarrow L_{n,m+1} \longrightarrow \alpha_p)$ is an epimorphism. Then $(X \longrightarrow L_{n,m+1} \longrightarrow L_{n,1})$ is an epimorphism by (10.3); consider the commutative diagram with exact rows:

$$\begin{array}{ccccccccc}
0 & \longrightarrow & X' & \longrightarrow & X & \longrightarrow & L_{n,1} & \longrightarrow & 0 \\
 & & \downarrow & & \downarrow & & \| & & \\
0 & \longrightarrow & L_{n,m} & \xrightarrow{V} & L_{n,m+1} & \xrightarrow{\rho} & L_{n,1} & \longrightarrow & 0 \\
 & & \downarrow{\scriptstyle\rho} & & \downarrow{\scriptstyle\rho} & & \downarrow{\scriptstyle\rho} & & \\
0 & \longrightarrow & \alpha_p & \xrightarrow{V} & L_{1,2} & \xrightarrow{\rho} & \alpha_p & \longrightarrow & 0.
\end{array}$$

As $L_{1,2} \longrightarrow \alpha_p$ is essential, $(X \longrightarrow L_{n,m+1} \longrightarrow L_{1,2})$ is an epimorphism. If $(X' \longrightarrow L_{n,m} \longrightarrow \alpha_p)$ would be zero, this would imply a factorization

$$(X \longrightarrow L_{1,2}) = (X \longrightarrow L_{n,1} \longrightarrow L_{1,2}),$$

a contradiction with (10.1), as $X \longrightarrow L_{1,2}$ is epimorphic. Thus it follows that $(X' \longrightarrow L_{n,m} \longrightarrow \alpha_p)$ is epimorphic, hence $X' = L_{n,m}$ by induction hypothesis, thus $X = L_{n,m+1}$, and the proposition is proved.

The proof of the following porposition is due to P.GABRIEL:

PROPOSITION (10.5): The exact sequence

$$0 \longrightarrow \alpha_p \xrightarrow{j} \underline{\underline{G}}_a \xrightarrow{F} \underline{\underline{G}}_a \longrightarrow 0$$

induces an epimorphism

$$j^*: \mathrm{Ext}(\underline{\underline{G}}_a, \underline{\underline{G}}_a) \longrightarrow \mathrm{Ext}(\alpha_p, \underline{\underline{G}}_a)$$

(and $\mathrm{Ext}(\alpha_p, \underline{\underline{G}}_a) \cong k$).

PROOF: An extension $X/\underline{\underline{G}}_a \cong \alpha_p$ admits a morphism-section; this follows using the criterion of GROTHENDIECK for smooth morphisms (cf. SGA, III, theorem 3.1) from the fact that $\underline{\underline{G}}_a$ is reduced (and hence $X \longrightarrow \alpha_p$ is smooth, cf. SGAD, VI.3.3.5). Thus the group $\mathrm{Ext}(\alpha_p, \underline{\underline{G}}_a)$ can be identified with $H^2_{reg}(\alpha_p, \underline{\underline{G}}_a)_s$ (notations of SERRE, this result is essentially contained in GA, VII.4, proposition 4a). Let us denote by $C^2(A,B)$, $A,B \in \underline{\underline{G}}$ the group of symmetric cocycles

$$f \in \mathrm{Mor}(A \times_k A, B), \quad \delta(f) = 0;$$

the group $H^2_{reg}(A,B)_s$ is the quotient of $C^2(A,B)$ by the subgroup of coboundaries (for notations, see GA, VII.4). The map

$$(j \times j)^* : C^2(\underline{\underline{G}}_a, \underline{\underline{G}}_a) \longrightarrow C^2(\alpha_p, \underline{\underline{G}}_a) \qquad (*)$$

is surjective: $f \in C^2(\alpha_p, \underline{\underline{G}}_a)$ is defined by a ringhomomorphism

$$\phi : k[X] \longrightarrow k[\tau] \otimes_k k[\tau] \quad , \quad \tau^p = 0;$$

the coefficients of the polynomial $\phi(X)$ (polynomial in the variables $\tau \otimes 1$ and $1 \otimes \tau$) we use to define a polynomial $\psi(X)$ in the variables $X \otimes 1$ and $1 \otimes X$. The corresponding ringhomomorphism

$$\psi : k[X] \longrightarrow k[X] \otimes_k k[X]$$

defines $g \in C^2(\underline{\underline{G}}_a, \underline{\underline{G}}_a)$, as follows immediately. Thus the map $(j \times j)^*$

in ($*$) is surjective, and the proposition is proved (the last statement follows by making explicit the calculation of LAZARD in the case $H^2_{reg}(\alpha_p, \mathbb{G}_a)_s$, or by calculating

$$\mathrm{Coker}(F^*: A_1 \longrightarrow A_1),$$

and using 9.1a).

QED

COROLLARY (10.6): For every $\phi \in A_1$, $\phi \neq 0$, the homomorphism

$$\phi^*: E^2(\mathbb{G}_a, \mathbb{G}_a) \longrightarrow E^2(\mathbb{G}_a, \mathbb{G}_a)$$

is injective (N.B. we shall prove that $E^2(\mathbb{G}_a, \mathbb{G}_a) = 0$).

PROOF: Any $\phi \in A_1$, diferent from zero, can be written in the form

$$\phi = \alpha_0 F^m (\alpha_1 - F) \times \ldots \times (\alpha_n - F) \quad \begin{cases} \alpha_0 \neq 0 \\ m \geq 0,\ n \geq 0. \end{cases}$$

Thus the result follows from (9.8) and (10.5).

QED

PROPOSITION (10.7): For all $n \geq 1$, $m \geq 1$,

$$E^2(W_m, W_n) = 0.$$

PROOF: Clearly it suffices to prove the case $n = 1 = m$. Let $\xi \in E^2(\mathbb{G}_a, \mathbb{G}_a)$ be defined by the exact sequence

$$0 \longrightarrow \mathbb{G}_a \longrightarrow X \longrightarrow Y \xrightarrow{\ f\ } \mathbb{G}_a \longrightarrow 0.$$

By (9.9) there exists a commutative diagram

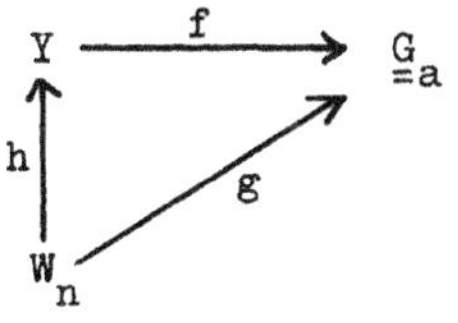

with epimorphic g. By (9.5a) there exists $\phi \in A_1$ such that

$g = \phi R^{n-1}$; as g is epimorphic, ϕ is not zero. The element $\phi^{*}(\xi) \in E^2(\underline{\underline{G}}_a, \underline{\underline{G}}_a)$ can be defined by the lower row of the exact, commutative diagram:

$$\begin{array}{ccccccccc}
0 & \longrightarrow & \underline{\underline{G}}_a & \longrightarrow & X & \longrightarrow & Y & \xrightarrow{f} & \underline{\underline{G}}_a & \longrightarrow & 0 \\
 & & \| & & \uparrow & & \uparrow h & & \uparrow \phi & & \\
0 & \longrightarrow & \underline{\underline{G}}_a & \longrightarrow & X' & \longrightarrow & W_n & \xrightarrow{R^{n-1}} & \underline{\underline{G}}_a & \longrightarrow & 0 .
\end{array}$$

The element $\phi^{*}(\xi)$ equals $\tau \cdot \alpha_1^{n-1}$:

$$0 \longrightarrow \underline{\underline{G}}_a \longrightarrow X' \longrightarrow W_{n-1} \longrightarrow 0 \qquad \tau \in \mathrm{Ext}(W_{n-1}, \underline{\underline{G}}_a),$$

$$0 \longrightarrow W_{n-1} \xrightarrow{V} W_n \xrightarrow{R^{n-1}} \underline{\underline{G}}_a \longrightarrow 0 \qquad \alpha_1^{n-1} .$$

From (9.1c) it follows that $\phi^{*}(\xi) = \tau \cdot \alpha_1^{n-1} = 0$. Hence $\xi = 0$ by (10.6), and the proposition is proved.

REMARK: It now follows easily that $E^2(\underline{\underline{G}}_a, \nu_p) \cong k$; this result (in the case of separable morphisms) was already obtained by ROSENLICHT (unpublished).

COROLLARY (10.8): $R : W_{n+1} \longrightarrow W_n$ induces the zero map

$$(R^{*}: E^2(W_n, \alpha_p) \longrightarrow E^2(W_{n+1}, \alpha_p)) = 0;$$

hence $E^2(W, \alpha_p) = 0$.

PROOF: The extensions $\underline{\underline{G}}_a / \alpha_p \cong \underline{\underline{G}}_a$ and $W_{n+1}/W_n \cong \underline{\underline{G}}_a$ yield a commutative, exact diagram

$$\begin{array}{ccccc}
A_1 & & & & \\
\downarrow & & & & \\
\operatorname{Ext}(W_n, \underline{\underline{G}}_a) & \longrightarrow & E^2(W_n, \alpha_p) & \longrightarrow & 0 = E^2(W_n, \underline{\underline{G}}_a) \\
\downarrow R^* & & \downarrow R^* & & \\
\operatorname{Ext}(W_{n+1}, \underline{\underline{G}}_a) & \longrightarrow & E^2(W_{n+1}, \alpha_p) & &
\end{array}$$

and the result follows by (9.1f).

QED

LEMMA (10.9): $F \in A_{n+1}$ induces the zero map

$$(F^* : \operatorname{Ext}(W_{n+1}, \alpha_p) \longrightarrow \operatorname{Ext}(W_{n+1}, \alpha_p)) = 0.$$

PROOF: The exact sequence $\underline{\underline{G}}_a/\alpha_p \cong \underline{\underline{G}}_a$ and $F \in A_{n+1}$ yield a commutative, exact diagram:

$$\begin{array}{ccccccc}
 & & \operatorname{Hom}(W_{n+1}, \underline{\underline{G}}_a) & \xrightarrow{\partial} & \operatorname{Ext}(W_{n+1}, \alpha_p) & \xrightarrow{i_*} & \operatorname{Ext}(W_{n+1}, \underline{\underline{G}}_a) \xrightarrow{F_*} \cdots \\
 & & \downarrow F^* & & \downarrow F^* & & \\
\operatorname{m}(W_{n+1}, \underline{\underline{G}}_a) & \xrightarrow{F_*} & \operatorname{Hom}(W_{n+1}, \underline{\underline{G}}_a) & \longrightarrow & \operatorname{Ext}(W_{n+1}, \alpha_p) . & &
\end{array}$$

By (9.2), ∂ is surjective, hence the lemma follows from (9.6).

QED

LEMMA (10.10): $\rho = RF : L_{n+1,m+1} \longrightarrow L_{n,m}$ induces the zero map

$$(\rho^* : \operatorname{Ext}(L_{n,m}, \alpha_p) \longrightarrow \operatorname{Ext}(L_{n+1,m+1}, \alpha_p)) = 0.$$

PROOF: By the definition of the groups $L_{i,j}$ we have a commutative diagram with exact rows:

$$\begin{array}{ccccc}
 & & \mathrm{Ext}(L_{n,m}, \alpha_p) & \xrightarrow{\partial} & E^2(W_n, \alpha_p) \\
 & & \downarrow R^* & & \downarrow R^* = 0 \\
\mathrm{Ext}(W_{n+1}, \alpha_p) & \longrightarrow & \mathrm{Ext}(L_{n,m+1}, \alpha_p) & \xrightarrow{\partial} & E^2(W_{n+1}, \alpha_p) \\
\downarrow F^* = 0 & & \downarrow F^* & & \\
\mathrm{Ext}(W_{n+1}, \alpha_p) & \longrightarrow & \mathrm{Ext}(L_{n+1,m+1}, \alpha_p). & &
\end{array}$$

The lemma follows by diagram chasing, using (10.8) and (10.9).

QED

<u>THEOREM</u> (10.11)(cf. GABRIEL, [16] , III, theorem 1): The object

$$L_{\infty,\infty} \longrightarrow \alpha_p$$

is the projective envelope of α_p in $\underline{P}$, and

$$0 \longrightarrow L_{\infty,\infty} \longrightarrow L_{\infty,\infty} \oplus L_{\infty,\infty} \longrightarrow L_{\infty,\infty} \longrightarrow \alpha_p \longrightarrow 0$$

is a projective resolution for α_p (the morphisms will be specified below). Hence the projective dimension of α_p equals two in $\underline{P}$.

<u>PROOF</u>: From (10.10) follows:

$$\mathrm{Ext}(L_{\infty,\infty}, \alpha_p) = \varinjlim \mathrm{Ext}(L_{n,m}, \alpha_p) = 0;$$

hence $L_{\infty,\infty}$ is projective in $\underline{P}_0$, and using (7.5) it follows that it is projective in $\underline{P}$. Taking into account (10.4), the first part of the theorem follows.

Consider the commutative, exact diagram:

$$\begin{array}{ccccccccc}
 & & & & & & 0 & & \\
 & & & & & & \downarrow & & \\
 & & & & & & L_{n,1} & & \\
 & & & & & & \downarrow & & \\
0 & \longrightarrow & L_{n,m} & \longrightarrow & W_n & \xrightarrow{F^m} & W_n & \longrightarrow & 0 \\
 & & \downarrow i & & \downarrow & & \downarrow F & & \\
0 & \longrightarrow & L_{n,m+1} & \longrightarrow & W_n & \xrightarrow{F^{m+1}} & W_n & \longrightarrow & 0 . \\
 & & & & & & \downarrow & & \\
 & & & & & & 0 & &
\end{array}$$

With the help of the snake lemma an exact sequence follows:

$$0 \longrightarrow L_{n,m} \xrightarrow{\;i\;} L_{n,m+1} \xrightarrow{\;F^m\;} L_{n,1} \longrightarrow 0.$$

From the exact sequence $W_{n+1}/W_n = \underline{\underline{G}}_a$ we deduce the exact sequence:

$$0 \longrightarrow L_{n,m} \xrightarrow{\;V\;} L_{n+1,m} \xrightarrow{\;R^n\;} L_{1,m} \longrightarrow 0.$$

Fitting together these sequences we obtain a commutative, exact diagram

$$\begin{array}{ccccccccc}
 & & 0 & & 0 & & 0 & & \\
 & & \downarrow & & \downarrow & & \downarrow & & \\
0 & \longrightarrow & L_{n,m} & \xrightarrow{V} & L_{n+1,m} & \xrightarrow{R^n} & L_{1,m} & \longrightarrow & 0 \\
 & & \downarrow{\scriptstyle i} & & \downarrow & & \downarrow & & \\
0 & \longrightarrow & L_{n,m+1} & \xrightarrow{V} & L_{n+1,m+1} & \xrightarrow{R^n} & L_{1,m+1} & \longrightarrow & 0 \\
 & & \downarrow{\scriptstyle F^m} & & \downarrow{\scriptstyle F^m} & & \downarrow & & \\
0 & \longrightarrow & L_{n,1} & \xrightarrow{V} & L_{n+1,1} & \xrightarrow{R^n} & \alpha_p & \longrightarrow & 0 . \\
 & & \downarrow & & \downarrow & & \downarrow & & \\
 & & 0 & & 0 & & 0 & &
\end{array}$$

Clearly all morphisms occurring in the diagram commute with R and with F, hence we can pass to the projective limit. As $\varprojlim$ is exact in $\underline{P}$, the resolution for α_p now follows directly from the following trivial:

<u>LEMMA</u> (10.12): Let

$$\begin{array}{ccccccccc}
 & & 0 & & 0 & & & & \\
 & & \downarrow & & \downarrow & & & & \\
0 & \longrightarrow & A_1 & \xrightarrow{\alpha} & A_2 & & & & \\
 & & \downarrow{\scriptstyle \beta} & & \downarrow{\scriptstyle \gamma} & & & & \\
0 & \longrightarrow & B_1 & \xrightarrow{\delta} & B_2 & & & & \\
 & & \downarrow & & \downarrow{\scriptstyle \lambda} & & & & \\
0 & \longrightarrow & C_1 & \longrightarrow & C_2 & \xrightarrow{\mu} & C_3 & \longrightarrow & 0 \\
 & & \downarrow & & \downarrow & & & & \\
 & & 0 & & 0 & & & &
\end{array}$$

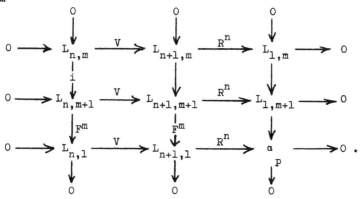

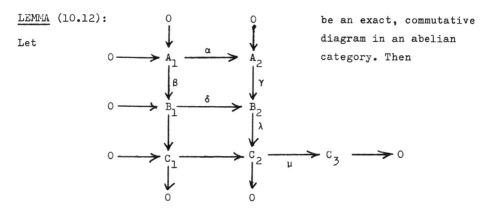

be an exact, commutative diagram in an abelian category. Then the sequence

$$0 \longrightarrow A_1 \xrightarrow{(\alpha,\beta)} A_2 \oplus B_1 \xrightarrow{(\gamma,-\delta)} B_2 \xrightarrow{\mu\lambda} C_3 \longrightarrow 0$$

is exact.

REMARK: CARTIER and GABRIEL have proved that unipotent group schemes can be regarded as certain modules over a ring A (see section 15 for the precise statement). With that result several parts of section 10 and section 11 can be straightened. In particular it follows from their result (in case the ground field is algebraically closed) that $\underline{\underline{G}}_a$ has injective dimension one in the category of unipotent group schemes (which implies 10.6), and that α_p has injective dimension two (which implies that its projective dimension in $\underline{P}_0$, and hence in $\underline{P}$, equals two).

II.11 The additive linear group

Let n be a natural number. We denote by Λ_n the set of all isogenies $f: G_f \longrightarrow W_n$, and we write Γ_n for the subset of Λ_n consisting of all isogenies of the form $f: W_n \longrightarrow W_n$. The inclusion $\Gamma_n \subset \Lambda_n$ induces a homomorphism

$$\phi_n : \overline{W}_n = \varprojlim_{f \in \Lambda_n} G_f \longrightarrow \varprojlim_{f \in \Gamma_n} W_n .$$

LEMMA (11.1): The homomorphisms ϕ_n are isomorphisms.

PROOF: If $f: G \longrightarrow W_n$ is an isogeny, there exists an isogeny $f' : W_n \longrightarrow G$ (cf. GA, VII.10, proposition 9). QED

COROLLARY (11.2): $(\pi_1(W_n))_{loc} = L_{n,\infty} \overset{def}{=} \varprojlim_m L_{n,m}$,

$(\pi_1(W))_{loc} = L_{\infty,\infty}$, and $\pi_1(W)$ is a projective object.

PROOF: Clearly $(\pi_1(W_n))_{loc}$ is the projective limit of $\mathrm{Ker}(\phi)$, for all purely inseparable $\phi \in A_n$. For such a ϕ there exists $\phi' \in A_n$ such that $\phi\phi' = F^m$ (cf. 6.2). This proves the first statement:

$$\begin{array}{ccccccccc}
0 & \longrightarrow & L_{n,m} & \longrightarrow & W_n & \xrightarrow{F^m} & W_n & \longrightarrow & 0 \\
 & & \downarrow & & \downarrow{\scriptstyle \phi'} & & \| & & \\
0 & \longrightarrow & \mathrm{Ker}(\phi) & \longrightarrow & W_n & \xrightarrow{\phi} & W_n & \longrightarrow & 0.
\end{array}$$

The second statement follows as π_1 commutes with projective limits:

$$\varprojlim_{n,m} L_{n,m} = \varprojlim_n (\varprojlim_m L_{n,m}) .$$

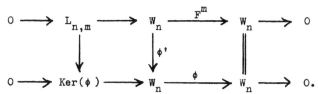

We know that $(\pi_1(W))_{loc} = L_{\infty,\infty}$ is a projective object (cf. 10.11), and for $(\pi_1(W))_{red}$ the arguments of GP, 8.4, apply.

QED

<u>COROLLARY</u> (11.3): a) $V : W_n \longrightarrow W_{n+1}$ induces a monomorphism

$$\overline{V} : \overline{W}_n \hookrightarrow \overline{W}_{n+1} ;$$

b) $\overline{V} : \overline{W} \hookrightarrow \overline{W}$ is a monomorphism.

<u>PROOF</u>: Consider the commutative, exact diagram:

$$\begin{array}{ccccc}
 & & 0 & & 0 \\
 & & \downarrow & & \downarrow \\
 & & \pi_1(W_n) & \xrightarrow{\alpha} & \pi_1(W_{n+1}) \\
 & & \downarrow & & \downarrow \\
 & & \overline{W}_n & \xrightarrow{\overline{V}} & \overline{W}_{n+1} \\
 & & \downarrow & & \downarrow \\
0 & \longrightarrow & W_n & \xrightarrow{V} & W_{n+1} \\
 & & \downarrow & & \downarrow \\
 & & 0 & & 0
\end{array}$$

The morphisms $V : L_{n,m} \longrightarrow L_{n+1,m}$ are monomorphic, hence (as $\varprojlim$ is exact in $\underline{P}$):

$$V : L_{n,\infty} \hookrightarrow L_{n+1,\infty}$$

is a monomorphism. By (11.2) we know that $L_{n,\infty} \cong (\pi_1(W_n))_{loc}$, and under this isomorphism α_{loc} corresponds to V; hence α_{loc} is a monomorphism. From GP, 8.4, corollary 1 it follows that α_{red} is also a monomorphism. Hence the first statement now follows by the five lemma. Statement (b) then follows, as $\varprojlim$ is exact in $\underline{P}$.

QED

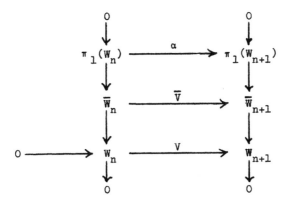

<u>PROPOSITION</u> (11.4): $\overline{R} : \overline{W}_{n+1} \longrightarrow \overline{W}_n$ induces the zero map

$$(\overline{R}^* : \operatorname{Ext}(\overline{W}_n, \underline{\underline{G}}_a) \longrightarrow \operatorname{Ext}(\overline{W}_{n+1}, \underline{\underline{G}}_a)) = 0.$$

<u>PROOF</u>: Let

$$\xi \in \operatorname{Ext}(\overline{W}_n, \underline{\underline{G}}_a) = \varinjlim_{f \in \Gamma_n} \operatorname{Ext}(W_n, \underline{\underline{G}}_a)$$

(cf. 11.1). Then we can give ξ by an element

$$\xi_f \in \operatorname{Ext}(G_f, \underline{\underline{G}}_a) \;,\quad G_f \cong W_n \,,\quad f : G_f \longrightarrow W_n \,.$$

By (9.7a) there exists a commutative diagram

$$\begin{array}{ccccc} W_{n+1} \cong H_g & \xrightarrow{\;R\;} & G_f \cong W_n \,. \\ \downarrow{\scriptstyle g} & & \downarrow{\scriptstyle f} \\ W_{n+1} & \xrightarrow{\;R\;} & W_n \end{array}$$

Hence $\overline{R}^*(\xi)$ can be represented by $R^*(\xi_f) \in \operatorname{Ext}(H_g, \underline{\underline{G}}_a)$. However, this element is zero (cf. 9.1f), thus the proposition is proved.

<u>COROLLARY</u> (11.5): $\operatorname{Ext}(\overline{W}, \underline{\underline{G}}_a) = 0$.

This follows as $\overline{W} = \varprojlim \overline{W}_n$.

<u>PROPOSITION</u> (11.6): $\operatorname{Ext}(\overline{W}_n, \alpha_p) = 0$.

<u>PROOF</u>: $\operatorname{Ext}(\overline{W}, \alpha_p) = \varinjlim_{f \in \Gamma_n} \operatorname{Ext}(G_f, \alpha_p)$ is represented by an element $\xi_f \in \operatorname{Ext}(G_f, \alpha_p)$ with $G_f = W_n$. As $F^*(\xi_f) = 0$ (cf. 10.9), the proposition is proved.

<u>COROLLARY</u> (11.7): $\overline{W}$ is a projective object in $\underline{P}$, the projective dimension of W equals one, and $\overline{W}$ is the projective envelope of $\underline{\underline{G}}_a$.

This follows from (11.5) and (11.6) (cf. GP, 8.6).

<u>COROLLARY</u> (11.8): The projective dimension of $\underline{\underline{G}}_a$ is two.

<u>PROOF</u>: By the exact sequence

$$0 \longrightarrow W \xrightarrow{\;V\;} W \longrightarrow \underline{\underline{G}}_a \longrightarrow 0$$

it follows that $dp(\underline{\underline{G}}_a) \leq 2$. As $E^2(\underline{\underline{G}}_a, \alpha_p) \neq 0$, in fact $dp(\underline{\underline{G}}_a) = 2$.

QED

<u>COROLLARY</u> (11.9): $\pi_2(\underline{\underline{G}}_a) = 0$ and the exact sequence

$$0 \longrightarrow \pi_1(W) \longrightarrow \pi_1(W) \oplus \overline{W} \longrightarrow \overline{W} \longrightarrow \underline{\underline{G}}_a \longrightarrow 0$$

is a projective resolution for $\underline{\underline{G}}_a$.

<u>PROOF</u>: From the exact sequence occurring in the proof of (11.8) follows the exact sequence

$$0 = \pi_2(W) \longrightarrow \pi_2(\underline{\underline{G}}_a) \longrightarrow \overline{W} \xhookrightarrow{\;\overline{V}\;} \overline{W}$$

(cf. GP, 6.2, proposition 5). Hence $\pi_2(\underline{\underline{G}}_a) = 0$ by (11.3b) and by $dp(W) = 1$ (cf. 11.7).

A projective resolution follows (cf. 10.12) from the commutative, exact diagram

$$\begin{array}{ccccccccc}
 & & 0 & & 0 & & & & \\
 & & \downarrow & & \downarrow & & & & \\
0 & \longrightarrow & \pi_1(W) & \longrightarrow & \pi_1(W) & & & & \\
 & & \downarrow & & \downarrow & & & & \\
0 & \longrightarrow & \overline{W} & \xrightarrow{\;\overline{V}\;} & \overline{W} & & & & \\
 & & \downarrow & & \downarrow & & & & \\
0 & \longrightarrow & W & \xrightarrow{\;V\;} & W & \longrightarrow & \underline{\underline{G}}_a & \longrightarrow & 0 \\
 & & \downarrow & & \downarrow & & & & \\
 & & 0 & & 0 & & & &
\end{array}$$

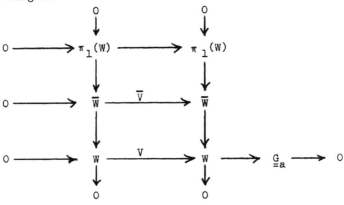

COROLLARY (11.10): $E^2(N, \underline{\underline{G}}_a) = 0$ for all $N \in \underline{P}_0$.

This follows, using (10.7), (11.8) and (8.3).

LEMMA (11.11): $F \in A_1$ induces the zero maps

$$(F_* : \mathrm{Hom}(\alpha_p, \underline{\underline{G}}_a) \longrightarrow \mathrm{Hom}(\alpha_p, \underline{\underline{G}}_a)) = 0$$

and

$$(F_* : \mathrm{Ext}(\alpha_p, \underline{\underline{G}}_a) \longrightarrow \mathrm{Ext}(\alpha_p, \underline{\underline{G}}_a)) = 0.$$

PROOF: The first statement is trivial. The second follows from

$$\mathrm{Im}(F_* : \mathrm{Ext}(\underline{\underline{G}}_a, \underline{\underline{G}}_a) \longrightarrow \mathrm{Ext}(\underline{\underline{G}}_a, \underline{\underline{G}}_a)) = \mathrm{Im}(F^* : \mathrm{Ext}(\underline{\underline{G}}_a, \underline{\underline{G}}_a) \longrightarrow \mathrm{Ext}(\underline{\underline{G}}_a, \underline{\underline{G}}_a))$$

and from the commutative, exact diagram

$$\begin{array}{ccccccc}
\mathrm{Ext}(\underline{\underline{G}}_a, \underline{\underline{G}}_a) & \xrightarrow{F^*} & \mathrm{Ext}(\underline{\underline{G}}_a, \underline{\underline{G}}_a) & \longrightarrow & \mathrm{Ext}(\alpha_p, \underline{\underline{G}}_a) & \longrightarrow & 0 \\
\downarrow{\scriptstyle F_*} & & \downarrow{\scriptstyle F_*} & & \downarrow{\scriptstyle F_*} & & \\
\mathrm{Ext}(\underline{\underline{G}}_a, \underline{\underline{G}}_a) & \xrightarrow{F^*} & \mathrm{Ext}(\underline{\underline{G}}_a, \underline{\underline{G}}_a) & \longrightarrow & \mathrm{Ext}(\alpha_p, \underline{\underline{G}}_a) & . &
\end{array}$$

QED

COROLLARY (11.12): $\mathrm{Hom}(\alpha_p, \underline{\underline{G}}_a) \cong k$, $\mathrm{Ext}(\alpha_p, \underline{\underline{G}}_a) \cong k$, $\mathrm{Ext}(\alpha_p, \alpha_p) \cong k^2$ and $E^2(\alpha_p, \alpha_p) \cong k$.

PROOF: The first two statements follow from the exact sequences

$$0 \longrightarrow A_1 \xrightarrow{F^*} A_1 \longrightarrow \mathrm{Hom}(\alpha_p, \underline{\underline{G}}_a) \longrightarrow 0$$

$$0 \longrightarrow \mathrm{Ext}(\underline{\underline{G}}_a, \underline{\underline{G}}_a) \xrightarrow{F^*} \mathrm{Ext}(\underline{\underline{G}}_a, \underline{\underline{G}}_a) \longrightarrow \mathrm{Ext}(\alpha_p, \underline{\underline{G}}_a) \longrightarrow 0.$$

The last two follow from the exact sequences

$$0 \longrightarrow \mathrm{Hom}(\alpha_p, \underline{\underline{G}}_a) \longrightarrow \mathrm{Ext}(\alpha_p, \alpha_p) \longrightarrow \mathrm{Ext}(\alpha_p, \underline{\underline{G}}_a) \longrightarrow 0$$

$$0 \longrightarrow \mathrm{Ext}(\alpha_p, \underline{\underline{G}}_a) \longrightarrow E^2(\alpha_p, \alpha_p) \longrightarrow 0 = E^2(\alpha_p, \underline{\underline{G}}_a).$$

QED

II.12 Abelian varieties

LEMMA (12.1): For any natural number $n \geq 1$, for any abelian variety B and for any $G \in \underline{G}$ the group $E^n(G,B)$ is a torsion group.

PROOF: $\xi \in E^n(G,B)$ can be written in the form $\xi = \xi_1 \cdot \xi_2$ $\xi_1 \in E^1(H,B)$ and $\xi_2 \in E^{n-1}(G,H)$, $H \in \underline{G}$. By the distributivity of the Yoneda-product it therefore suffices to prove the case $n = 1$. By exactness of the sequence

$$\mathrm{Ext}(H/CR(H),B) \longrightarrow \mathrm{Ext}(H,B) \longrightarrow \mathrm{Ext}(CR(H),B)$$

it suffices to consider the cases $H \in \underline{G}_{CR}$ and $H \in \underline{N}$. In the first case it is a result of SERRE (cf. [35] , 5.3, lemma 7 = GP, 7.4, proposition 4). Using (3.1) the second case follows, as $q \cdot 1_H = 0$ for $H \in \underline{N}$ and suitable q.

QED

LEMMA (12.2): For any abelian variety A and any natural number q,

$$E^2(A, \mu_q) = 0 = E^2(A, \nu_p).$$

PROOF: The sequence

$$0 \longrightarrow E^2(A, \mu_q) \longrightarrow E^2({}_qA, \mu_q)$$

is exact. On the other hand ${}_qA \in \underline{N}$ implies $E^2({}_qA, \mu_q) = 0$. Analogous proof in the case ν_p.

QED

PROPOSITION (12.3): $E^2(-, \underline{\underline{G}}_m) = 0$.

PROOF: It suffices to prove $E^2(X, \underline{\underline{G}}_m) = 0$ for every elementary $X \in \underline{G}$. First: $dp(\underline{\underline{G}}_m) = 1 = dp(\mu_q) = dp(\nu_p)$. Further: $\mathrm{Hom}(L_{\infty,\infty}, \underline{\underline{G}}_m) = 0$, hence $E^2(\alpha_p, \underline{\underline{G}}_m) = 0$ by (10.11), and $\mathrm{Hom}(\pi_1(W), \underline{\underline{G}}_m) = 0$, thus $E^2(\underline{\underline{G}}_a, \underline{\underline{G}}_m) = 0$ by (11.9). Hence the case of an abelian variety remains.

We first prove that $E^2(A, \underline{\underline{G}}_m)$ is a torsion group for every abelian variety A (for a different proof, compare section 15). Let

$E^2(A,\underline{\underline{G}}_m)$ be defined by an exact sequence

$$0 \longrightarrow \underline{\underline{G}}_m \longrightarrow Y_1 \longrightarrow Y_2 \longrightarrow A \longrightarrow 0.$$
$$Y_1 \twoheadrightarrow K \hookrightarrow Y_2$$

Let K_1 be the maximal connected linear subgroup of K_{red}, and $L = K/K_1$. We define $\alpha \in \mathrm{Ext}(A,L)$ by the lower row of the commutative, exact diagram

$$\begin{array}{ccccccccc} 0 & \longrightarrow & K & \longrightarrow & Y_2 & \longrightarrow & A & \longrightarrow & 0 \\ & & \downarrow & & \downarrow & & \| & & \\ 0 & \longrightarrow & L & \longrightarrow & Z & \longrightarrow & A & \longrightarrow & 0. \end{array}$$

As L_{red} is an abelian variety, there exists a number n such that $n\cdot\alpha = 0$ (cf. 12.1). As $\mathrm{Ext}(K_1,\underline{\underline{G}}_m) = 0$, it follows that $n\cdot\xi = 0$.

Suppose $\xi \in E^2(A,\underline{\underline{G}}_m)$, A an abelian variety and $n\cdot\xi = 0$. Exactness of the sequence

$$0 = E^2(A,{}_n(\underline{\underline{G}}_m)) \longrightarrow E^2(A,\underline{\underline{G}}_m) \xrightarrow{\ n\ } E^2(A,\underline{\underline{G}}_m)$$

(cf. 12.2) proves $\xi = 0$. QED

<u>LEMMA</u> (12.4): For every abelian variety B and every affine group scheme G, $E^2(G,B) = 0$.

<u>PROOF</u>: The cases $E^2(\alpha_p,B) = 0 = E^2(\underline{\underline{G}}_a,B)$ follow by the exact sequence

$$E^2(G,B) \xrightarrow{\times p = 0} E^2(G,B) \longrightarrow E^3(G,{}_pB) = 0.$$

As $dp(\mu_q) = 1 = dp(\nu_p)$ it follows that $E^2(X,B) = 0$ for all $X \in \underline{N}$; the case $E^2(\underline{\underline{G}}_m,B) = 0$ follows from (8.3). QED

<u>PROPOSITION</u> (12.5): For every abelian variety **A**, $dp(A) \leq 2$.

<u>PROOF</u>: It suffices to prove $E^3(A,X) = 0$ for all elementary $X \in \underline{G}$.

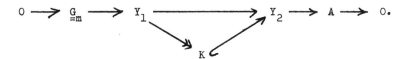

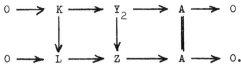

From (12.3) the case $E^3(A,\underline{\underline{G}}_m) = 0$ follows. In the cases $X = \underline{\underline{G}}_a$, $X = \mu_p$, $X = \nu_p$ (respectively $X = \mu_q$) the exact sequence

$$E^3(A,X) \xrightarrow{\times p = 0} E^3(A,X) \longrightarrow E^3({}_pA,X)$$

(respectively $0 \longrightarrow E^3(A,X) \longrightarrow E^3({}_qA,X)$) together with $dp({}_pA) \leq 2$ (respectively $dp({}_qA) \leq 2$) yield $E^3(A,X) = 0$.

Let B be an abelian variety. For any natural number q, the sequence

$$E^2({}_qA,B) \longrightarrow E^3(A,B) \xrightarrow{\times q} E^3(A,B)$$

is exact. As ${}_qA \in \underline{N}$, $E^2({}_qA,B) = 0$ (cf. 12.4). As $E^3(A,B)$ is a torsion group (cf. 12.1), this proves $E^3(A,B) = 0$ in case B is an abelian variety. Hence all cases are settled, and the proposition is proved.

For any $G \in \underline{G}$ the group $Hom(\alpha_p,G)$ has the structure of a right $Hom(\alpha_p,\alpha_p)$ - module. As $Hom(\alpha_p,\alpha_p) \cong k$ (ring-isomorphism), the group $Hom(\alpha_p,G)$ has in a natural way the structure of a k-vectorspace. It is easy to see that for every $G \in \underline{G}$ the dimension is finite: $Hom(\alpha_p,G) \cong Hom(\alpha_p,\underline{\underline{I}}^1(G))$, and use $Hom(\alpha_p,\alpha_p) \cong k$.

<u>DEFINITION</u>: For any $G \in \underline{G}$ we define

$$\tau(G) = \dim_k Hom(\alpha_p,G).$$

<u>LEMMA</u> (12.6): For any $G \in \underline{G}$, $G = G_{red}$, $\tau(G) + \sigma(G) \leq \dim G$.

<u>PROOF</u>: This follows from the fact that $(\underline{\underline{I}}^1(G))_{11}$ is successive extension of precisely $\dim G - \sigma(G)$ copies of α_p.

<u>EXAMPLES</u>: $\tau(\underline{\underline{G}}_m) = 0$; $\tau(\underline{\underline{G}}_a) = 1 = \tau(W_n)$ for all n; if $\tau(A) = \dim A$ and $A = A_{red}$ then $\sigma(A) = 0$. Note that $\tau(-)$ is not in-

variant under isogenies (compare section 15).

LEMMA (12.7): Let $N \in \underline{N}_{11}$, $N \neq 0$. Then $\mathrm{Hom}(\alpha_p, N) \neq 0$, and $E^2(N, \alpha_p) \neq 0$. If $G \in \underline{G}$, $\sigma(G) < \dim G$, then $\tau(G) \neq 0$.

PROOF: N is successive extension of copies of α_p. Hence the first statement follows from $\mathrm{Hom}(\alpha_p, \alpha_p) \neq 0$ and the second follows from $E^2(\alpha_p, \alpha_p) \neq 0$ (cf. 11.12) and from $E^3(-,-) = 0$. If $\sigma(G) < \dim G$, $(\underline{\underline{I}}^1(G))_{11} \neq 0$, hence the last assertion is clear.

QED

LEMMA (12.8): For any abelian variety A, $E^2(A, \underline{\underline{G}}_a) = 0$.

PROOF: Use the exact sequence (and use (11.10)):

$$E^2(A, \underline{\underline{G}}_a) \xrightarrow{\times p = 0} E^2(A, \underline{\underline{G}}_a) \xrightarrow{\sim} E^2({}_pA, \underline{\underline{G}}_a) = 0.$$

QED

PROPOSITION (12.9): Let $A \neq 0$ be an abelian variety. If $\tau(A) \neq 0$, $dp(A) = 2$, and if $\tau(A) = 0$, $dp(A) = 1$.

PROOF: Assume $\tau(A) \neq 0$. In that case $(\underline{\underline{I}}^1(A))_{11} \neq 0$, hence $({}_pA)_{11} \neq 0$, as ${}_pA \supset \underline{\underline{I}}^1(A)$. The exact sequence

$$(*) \qquad 0 \longrightarrow E^2(A, \alpha_p) \longrightarrow E^2({}_pA, \alpha_p) \longrightarrow E^3(A, \alpha_p) = 0$$

proves:

$$E^2(A, \alpha_p) \xrightarrow{\sim} E^2({}_pA, \alpha_p) \neq 0$$

(cf. 12.7); hence $dp(A) \geq 2$. By (12.5) we obtain $dp(A) = 2$.

Assume $\tau(A) = 0$. In that case $({}_pA)_{11} = 0$ by (12.7), hence $E^2(A, \alpha_p) = 0$ by exactness of the sequence $(*)$. We know that $E^2(A, X) = 0$ in the cases $X = \underline{\underline{G}}_a$ (cf. 12.8), $X = \underline{\underline{G}}_m$ (cf. 12.3),

$X = \mu_q$ and $X = \nu_p$ (cf. 12.2).

Let q be a positive integer, and let B be an abelian variety. Then $\tau(A) = 0$ implies $\mathrm{Ext}({}_qA, B) = 0$ ($({}_qA)_{11} = 0$, use (8.3) and the exact sequence $0 \longrightarrow \mathrm{Ext}(\mu_q, B) \longrightarrow E^2(\mu_q, {}_qB) = 0$). As $E^2(A,B)$ is a torsion group (cf. 12.1) exactness of the sequence

$$0 = \mathrm{Ext}({}_qA, B) \longrightarrow E^2(A,B) \xrightarrow{\times q} E^2(A,B)$$

proves $E^2(A,B) = 0$ in case $\tau(A) = 0$. Thus $dp(A) \leq 1$, hence it follows that $dp(A) = 1$ as for example $\mathrm{Ext}(A, \underline{\underline{G}}_a) \neq 0$.

QED

<u>REMARK</u>: If $G \in \underline{G}$, $G \neq 0$, then $\tau(G) \neq 0$ if and only if $dp(G) = 2$ and $\tau(G) = 0$ if and only if $dp(G) = 1$. Let $B \neq 0$ be an abelian variety; then $\tau(B) = 0$ implies that the injective dimension of B is one.

<u>PROPOSITION</u> (12.10): Let A and B be isogeneous abelian varieties. Then $\tau(A) = 0$ if and only if $\tau(B) = 0$. Hence in this case $dp(A) = dp(B)$.

<u>PROOF</u>: $\tau(A) = 0$ if and only if $\sigma(A) = \dim A = \dim B$. As $\sigma(A) = \sigma(B)$ (cf. 8.6) the first statement follows. The second results from (12.9), and the proposition is proved.

Let A be an abelian variety. The kernel of $p \cdot 1_A : A \longrightarrow A$ is finite. We define the integer $s(A)$ by:

$$|({}_pA)_k| = p^{s(A)} = |\mathrm{Hom}(\nu_p, A)| .$$

We remark that the separable degree of $p \cdot 1_A$ equals $p^{s(A)}$.

<u>LEMMA</u> (12.11): If A and B are isogeneous abelian varieties, $s(A) = s(B)$.

<u>PROOF</u>: Let $f: A \longrightarrow B$ be an elementary isogeny, i.e. $\mathrm{Ker}(f) = N$ is an elementary, finite group scheme. As $E^2(\nu_p, {}_pA) = 0$ it follows that $\mathrm{Ext}(\nu_p, A) = 0$. Hence we can use the exact sequence

$$0 \longrightarrow \mathrm{Hom}(\nu_p, N) \longrightarrow \mathrm{Hom}(\nu_p, A) \longrightarrow$$
$$\longrightarrow \mathrm{Hom}(\nu_p, B) \longrightarrow \mathrm{Ext}(\nu_p, N) \longrightarrow 0.$$

QED

<u>PROPOSITION</u> (12.12): For an abelian variety A it holds

$$\sigma(A) = s(A).$$

<u>PROOF</u>: Clearly

$$|\mathrm{Hom}({}_pA, \underline{\underline{G}}_m)| = p^{\sigma(A)}.$$

Let A^t denote the dual abelian variety (cf. section 5). The Weil--Barsotti formula reads:

$$\mathrm{Ext}(A, \underline{\underline{G}}_m) \xrightarrow{\sim} \mathrm{Mor}(\mathrm{Spec}(k), A^t) = A^t_k$$

(cf. GA, VII.16, theorem 6). We obtain a commutative, exact diagram

$$\begin{array}{ccccccc}
0 \longrightarrow & {}_p(A^t_k) & \longrightarrow & A^t_k & \xrightarrow{\times p} & A^t_k & \\
 & & & \uparrow \wr & & \uparrow \wr & \\
0 \longrightarrow & \mathrm{Hom}({}_pA, \underline{\underline{G}}_m) & \longrightarrow & \mathrm{Ext}(A, \underline{\underline{G}}_m) & \xrightarrow{\times p} & \mathrm{Ext}(A, \underline{\underline{G}}_m), &
\end{array}$$

which proves

$$p^{s(A^t)} = |{}_p(A^t_k)| = |\mathrm{Hom}({}_pA, \underline{\underline{G}}_m)| = p^{\sigma(A)}.$$

As A and A^t are isogeneous (for example, cf. [22], IV.4, theorem 10), this proves (cf. 12.11):

$$\sigma(A) = s(A^t) = s(A).$$

QED

REMARK: This proposition is a special case of a duality theory of abelian schemes, which will be studied in section 19.

LEMMA (12.13): Let A be an abelian variety. Then

$$\dim_k \operatorname{Ext}(A, \alpha_p) = \tau(A^t) = \dim_k E^2(A, \alpha_p).$$

PROOF: Exactness of the sequence

$$0 \longrightarrow \operatorname{Hom}({}_pA, \alpha_p) \longrightarrow \operatorname{Ext}(A, \alpha_p) \xrightarrow{\times p=0} \operatorname{Ext}(A, \alpha_p)$$

yields

$$\dim_k \operatorname{Ext}(A, \alpha_p) \cong \dim_k \operatorname{Hom}({}_pA, \alpha_p).$$

It follows from a theorem we shall prove in section 19, that $({}_pA)^D \cong {}_p(A^t)$, hence

$$\dim_k \operatorname{Ext}(A, \alpha_p) = \dim_k \operatorname{Hom}(\alpha_p, {}_p(A^t)) = \tau(A^t).$$

From the exact sequence (of k-vectorspaces)

$$0 \longrightarrow \operatorname{Ext}(A, \alpha_p) \longrightarrow \operatorname{Ext}(A, \underline{\underline{G}}_a) \xrightarrow{F_*} \operatorname{Ext}(A, \underline{\underline{G}}_a) \longrightarrow E^2(A, \alpha_p) \longrightarrow 0$$

it follows that

$$\dim_k \operatorname{Ext}(A, \alpha_p) = \dim_k E^2(A, \alpha_p).$$

QED

For more information concerning abelian varieties, we refer to section 15.

II.13 A conjecture of SERRE

Let A be an abelian variety. One can prove that $\text{Ext}(\underline{\underline{G}}_a, A)$ is a finite dimensional (right) vectorspace over k, in fact ($\dim A = n$):

$$\dim_k \text{Ext}(\underline{\underline{G}}_a, A) \leq 2n .$$

In a letter to me (6/2/1963) Professor Serre conjectured:

$$\dim_k \text{Ext}(\underline{\underline{G}}_a, A) = n$$

(remind that the characteristic of k is different from zero). In this section we prove this to be true.

It was communicated to me that MIYANISHI (Kyoto University) has obtained the same result (cf. [25]). Making use of this, MATSUMURA and MIYANISHI have defined a nice duality between $\text{Ext}(A, \underline{\underline{G}}_a)$ and $\text{Ext}(\underline{\underline{G}}_a, A)$ (hence $\text{Ext}(\underline{\underline{G}}_a, A)$ can be identified with the tangent space to A at the origin in a functorial manner)(cf. [24]).

THEOREM (13.1): For any abelian variety A it holds

$$\dim_k \text{Ext}(\underline{\underline{G}}_a, A) = \dim A .$$

The proof starts as follows. Consider the exact sequence

$$0 \longrightarrow {}_pA \longrightarrow A \xrightarrow{p \cdot 1_A} A \longrightarrow 0.$$

This yields an exact sequence

$$0 \longrightarrow \text{Ext}(\underline{\underline{G}}_a, {}_pA) \longrightarrow \text{Ext}(\underline{\underline{G}}_a, A) \xrightarrow{\times p} \text{Ext}(\underline{\underline{G}}_a, A)$$

(cf. 3.1), and as $p \cdot 1_{\underline{\underline{G}}_a} = 0$, this proves that

$$\text{Ext}(\underline{\underline{G}}_a, {}_pA) \xrightarrow{\sim} \text{Ext}(\underline{\underline{G}}_a, A)$$

is an isomorphism. Thus

$$\mathrm{Ext}(\underline{\underline{G}}_a, A) = \mathrm{Ext}(\underline{\underline{G}}_a, ({}_pA)_{rl}) \oplus \mathrm{Ext}(\underline{\underline{G}}_a, ({}_pA)_{ll})$$

(in fact this is the Jordan reduction of $\mathrm{Ext}(\underline{\underline{G}}_a, {}_pA)$ with respect to F^*). As $({}_pA)_{rl} = (\nu_p)^{\sigma(A)}$,

$$\dim_k \mathrm{Ext}(\underline{\underline{G}}_a, ({}_pA)_{rl}) = \sigma(A),$$

and we have to show

$$\dim_k \mathrm{Ext}(\underline{\underline{G}}_a, ({}_pA)_{ll}) = \dim A - \sigma(A).$$

Consider the exact, commutative diagram (i is some positive integer):

$$\begin{array}{ccccccccc}
 & & & & 0 & & 0 & & \\
 & & & & \downarrow & & \downarrow & & \\
0 & \longrightarrow & \alpha_p & \longrightarrow & L_{1,i+1} & \longrightarrow & L_{1,i} & \longrightarrow & 0 \\
 & & \| & & \downarrow & & \downarrow & & \\
0 & \longrightarrow & \alpha_p & \longrightarrow & \underline{\underline{G}}_a & \xrightarrow{F} & \underline{\underline{G}}_a & \longrightarrow & 0 \\
 & & & & \downarrow{\scriptstyle F^{i+1}} & & \downarrow{\scriptstyle F^{i}} & & \\
 & & & & \underline{\underline{G}}_a & = & \underline{\underline{G}}_a & & \\
 & & & & \downarrow & & \downarrow & & \\
 & & & & 0 & & 0 & &
\end{array}$$

We define $L_{1,\infty} = \varprojlim L_{1,i}$.

<u>LEMMA</u> (13.2): $\mathrm{Ext}(L_{1,\infty}, \alpha_p) \xrightarrow{\sim} E^2(\underline{\underline{G}}_a, \alpha_p)$ is an isomorphism.

<u>PROOF</u>: The Hom-Ext exact sequence defined by the last row of the diagram splices:

$$0 \longrightarrow \mathrm{Hom}(L_{1,i}, \alpha_p) \longrightarrow \mathrm{Ext}(\underline{\underline{G}}_a, \alpha_p) \longrightarrow 0$$

$$0 \longrightarrow \mathrm{Ext}(\underline{\underline{G}}_a, \alpha_p) \longrightarrow \mathrm{Ext}(L_{1,i}, \alpha_p) \longrightarrow E^2(\underline{\underline{G}}_a, \alpha_p) \longrightarrow 0$$

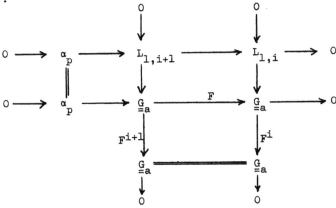

$$0 \longrightarrow E^2(\underline{\underline{G}}_a, \alpha_p) \longrightarrow E^2(L_{1,i}, \alpha_p) \longrightarrow 0$$

use dim $\mathrm{Ext}(\underline{\underline{G}}_a, \alpha_p) = 1 = \dim E^2(\underline{\underline{G}}_a, \alpha_p)$). As F^* is zero on $\mathrm{Ext}(\underline{\underline{G}}_a, \alpha_p)$ (cf. 10.9) and as $\varinjlim$ is exact in the category of abelian groups, the lemma follows.

<u>COROLLARY</u> (13.3): Let M be a finite group scheme. Then

$$\mathrm{Hom}(L_{1,\infty}, M) \xrightarrow{\sim} \mathrm{Ext}(\underline{\underline{G}}_a, M_{loc})$$

is an isomorphism.

<u>PROOF</u>: This is clear in case $M_{loc} = \alpha_p$ as F^* is zero on $\mathrm{Ext}(\underline{\underline{G}}_a, \alpha_p)$. We use induction on the length of a composition sequence of M_{11}. Suppose $M_{11} = N$, $\alpha_p \subset N$, $N/\alpha_p = N'$ such that we have proved

$$\mathrm{Hom}(L_{1,\infty}, N') \xrightarrow{\sim} \mathrm{Ext}(\underline{\underline{G}}_a, N')$$

to be an isomorphism. Then we conclude by diagram chasing:

$$\begin{array}{ccccccccc}
0 \rightarrow & \mathrm{Hom}(L_{1,\infty}, \alpha_p) & \rightarrow & \mathrm{Hom}(L_{1,\infty}, N) & \rightarrow & \mathrm{Hom}(L_{1,\infty}, N') & \rightarrow & \mathrm{Ext}(L_{1,\infty}, \alpha_p) \\
& \downarrow \cong & & \downarrow & & \downarrow \cong & & (13.2) \downarrow \cong \\
0 \rightarrow & \mathrm{Ext}(\underline{\underline{G}}_a, \alpha_p) & \rightarrow & \mathrm{Ext}(\underline{\underline{G}}_a, N) & \rightarrow & \mathrm{Ext}(\underline{\underline{G}}_a, N') & \rightarrow & E^2(\underline{\underline{G}}_a, \alpha_p).
\end{array}$$

QED

We are going to use the following facts:

<u>(13.4)</u>: The dual of $L_{1,i}$ is isomorphic to $L_{i,1}$ (cf. [16], I.5; the general relation $L^D_{i,j} = L_{j,i}$ also holds, cf. section 10).

<u>(13.5)</u>: Let X be a projective, normal, reduced k-algebraic scheme. We define

$$\nu(X) = \sup_{i \to \infty} \dim_k H^0(X, \underline{W}_i / F\underline{W}_i)$$

(cf. [37], section 7; $\underline{W}_i$ denotes the sheaf of Witt-vectors of length i on X). If X is an abelian variety, then

$$\nu(X) = \dim X - \sigma(X).$$

This follows from

$$rg_{\Lambda} H^1(X,\underline{W}) = \dim X + \nu(X)$$

(cf. [37], 7, proposition 4, use [36], 2, theorem 2), and from

$$rg_{\Lambda} H^1(X,\underline{W}) = 2.\dim X - \sigma(X)$$

(cf. [36], 11, corollary 1 of theorem 6).

We now conclude the <u>proof of the theorem</u>. For any positive integer j, consider the exact sequence

$$0 \longrightarrow L_{j,1} \longrightarrow W_j \xrightarrow{F} W_j \longrightarrow 0.$$

As

$$\mathrm{Ext}(A^t,W_j) \cong H^1(A^t,\underline{W}_j)$$

(cf. GA, VII.18, theorem 8), we deduce an exact sequence

$$0 \longrightarrow \mathrm{Ext}(A^t,L_{j,1}) \longrightarrow H^1(A^t,\underline{W}_j) \xrightarrow{F_*} H^1(A^t,\underline{W}_j).$$

By exactness of the sequence

$$0 \longrightarrow H^0(A^t,\underline{W}_j/F\underline{W}_j) \longrightarrow H^1(A^t,\underline{W}_j) \xrightarrow{F_*} H^1(A^t,\underline{W}_j)$$

(cf. [37], page 36), this proves

$$\mathrm{Ext}(A^t,L_{j,1}) \cong H^0(A^t,\underline{W}_j/F\underline{W}_j);$$

in particular (cf. 13.5):

$$\sup_{j \to \infty} \dim_k \mathrm{Ext}(A^t,L_{j,1}) = \dim(A^t) - \sigma \qquad (1)$$

($\sigma(A) = \sigma = \sigma(A^t)$). As multiplication by p on $L_{j,1}$ is the zero homomorphism (multiplication by p on $L_{1,j}$ is zero), we deduce an isomorphism

$$\mathrm{Hom}({}_p(A^t), L_{j,1}) \xrightarrow{\sim} \mathrm{Ext}(A^t, L_{j,1}) \qquad (2)$$

(use the exact sequence $A^t/{}_pA^t = A^t$). We shall later prove

$$({}_pA)^D \cong {}_p(A^t)$$

(cf. section 19). Thus we obtain:

$$\mathrm{Hom}(L_{1,j}, {}_pA) = \mathrm{Hom}(({}_pA)^D, L^D_{1,j}) =$$

$$= \mathrm{Hom}({}_p(A^t), L_{j,1}) \overset{(2)}{=\!=} \mathrm{Ext}(A^t, L_{j,1}).$$

Thus:

$$\dim_k \mathrm{Ext}(\underline{\underline{G}}_a, ({}_pA)_{11}) \overset{(13.3)}{=\!=\!=} \dim_k \mathrm{Hom}(L_{1,\infty}, {}_pA) =$$

$$= \sup_{j \to \infty} \dim_k \mathrm{Ext}(A^t, L_{j,1}) \overset{(1)}{=\!=} \dim(A^t) - \sigma = \dim(A) - \sigma(A),$$

and the theorem is proved by what is said on page 13-2.

II.14 The homological dimension of $\underline{G}$ and extensions of elementary groups

In this section we state the results we obtained in the sections before. The main result of GP holds also in our setting:

THEOREM (14.1): Let k be an algebraically closed field of characteristic $p \neq 0$. The category $\underline{G} = \underline{G}_k$ of commutative k-algebraic group schemes has homological dimension 2.

In fact, we proved that of all elementary groups the projective dimension is at most two: μ_q and ν_p: section 2 and (7.5); $\underline{\underline{G}}_m$: (8.3); α_p : (10.11); $\underline{\underline{G}}_a$: (11.9); abelian varieties: (12.5).

THEOREM (14.2): For all $G \in \underline{P}$ and every $i \geq 2$,

$$\pi_i(G) = 0.$$

It suffices to prove this in case $i = 2$. This is clear for $G \in \underline{P}_0$ and $G = \underline{\underline{G}}_m$. The case $G = \underline{\underline{G}}_a$ was settled in (11.9). The case of an abelian variety was proved in (7.3).

COROLLARY (11.3): The functor $G \longmapsto \overline{G}$ is exact.

We make explicit some calculations:

LEMMA (14.4): Let B be an abelian variety; then $\dim_k \mathrm{Ext}(\alpha_p, B) = \tau(B)$.

PROOF: Calculate dimensions in the exact sequence

$$0 \longrightarrow \mathrm{Hom}(\alpha_p, B) \longrightarrow \mathrm{Ext}(\underline{\underline{G}}_a, B) \xrightarrow{F^*} \mathrm{Ext}(\underline{\underline{G}}_a, B) \longrightarrow \mathrm{Ext}(\alpha_p, A) \longrightarrow 0.$$

QED

$(r,q) = 1 = (q,p) = (r,p)$
A and B abelian varieties
$\dim A = m$, $\dim B = n$

X \ Y →	projective dim	$\underline{\underline{G}}_a$	$\underline{\underline{G}}_m$	B	μ_q	ν_p	α_p	μ_p
Hom(X,Y) — injective dimension		1	1	1 or 2	1	1	2	1
$\underline{\underline{G}}_a$	2	A_1	0	0	0	0	0	0
$\underline{\underline{G}}_m$	1	0	$\mathbb{Z}$	0	0	0	0	0
A	1 2	0	0	[41], X.70	0	0	0	0
μ_r	1	0	$\mathbb{Z}/r\,\mathbb{Z}$	$(\mathbb{Z}/r\,\mathbb{Z})^{2n}$	0	0	0	0
ν_p	1	k^+	0	$(\mathbb{Z}/p\mathbb{Z})^{s(B)}$	0	$\mathbb{Z}/p\,\mathbb{Z}$	0	0
α_p	2	k^+, 11.12	0	$(k^+)^{\tau(B)}$	0	0	k	0
μ_p	1	0	$\mathbb{Z}/p\mathbb{Z}$	$(\mathbb{Z}/p\mathbb{Z})^{\sigma(B)}$	0	0	0	$\mathbb{Z}/p\mathbb{Z}$
Ext(X,Y)								
$\underline{\underline{G}}_a$		A_1^+	0 [21]	$(k^+)^n$ 13.1	0	k^+	k^+	0
$\underline{\underline{G}}_m$		0, [21]	0	$\mathrm{Hom}(\pi_1 \underline{\underline{G}}_m, B)$	$\mathbb{Z}/q\,\mathbb{Z}$	0	0	$\mathbb{Z}/p\,\mathbb{Z}$
A		$H^1(A, \underline{O}_A)$ GA, VII.17	$A_k^t = H^1(A, \underline{O}_A^*)_0$ GA, VII.16	torsion $\neq 0$, 14.7	$(\mathbb{Z}/q\mathbb{Z})^{2m}$	$(\mathbb{Z}/p\mathbb{Z})^{sA}$	$(k^+)^{\tau(A^t)}$	$(\mathbb{Z}/p\mathbb{Z})^{s(A)}$
μ_r		0	0	0	0	0	0	0
ν_p		k^+, 9.8	0	0	0	$\mathbb{Z}/p\,\mathbb{Z}$	0	0
α_p		k^+, 10.5	0	$(k^+)^{\tau(B)}$, 14.4	0	0	$(k^+)^2$ 11.12	0
μ_p		0, 8.5	0, 8.5	0, 8.5	0	0	0	$\mathbb{Z}/p\,\mathbb{Z}$
$E^2(X,Y)$								
$\underline{\underline{G}}_a$		0, 10.7	0, 12.3	0, 12.4	0, 14.5	k^+, 10	k^+, 14.6	0
A		0, 12.8	0, 12.3	Compare p. 15-13	0, 12.2	0, 12.2	$(k^+)^{\tau(A^t)}$ 12.13	0 12.2
α_p		0, use 10.7, 11.8	0, 12.3	0, 12.4	0	0	k^+, 8.12	0

LEMMA (14.5): $E^2(\underline{\underline{G}}_a, \mu_q) = 0$.

This follows from $\mathrm{Ext}(\underline{\underline{G}}_a, \underline{\underline{G}}_m) = 0 = E^2(\underline{\underline{G}}_a, \underline{\underline{G}}_m)$.

LEMMA (14.6): $\dim_k E^2(\underline{\underline{G}}_a, \alpha_p) = 1$

PROOF: In the exact sequence

$$\mathrm{Ext}(\underline{\underline{G}}_a, \underline{\underline{G}}_a) \xrightarrow{F_*} \mathrm{Ext}(\underline{\underline{G}}_a, \underline{\underline{G}}_a) \longrightarrow E^2(\underline{\underline{G}}_a, \alpha_p) \longrightarrow 0$$

the cokernel of F_* has dimension 1. QED

LEMMA (14.7): Let A and B be abelian varieties. Then $\mathrm{Ext}(A,B) \neq 0$.

PROOF: The group Ext(A,B) has elements with arbitrarily high order. This implies that $\mathrm{Ext}(A,B) = 0$ if and only if $\mathrm{Ext}(A',B') = 0$, in case A' is isogeneous with A, and B' is isogeneous with B. Moreover every abelian variety is isogeneous with a direct sum of elementary abelian varieties; thus in order to prove the lemma, it suffices to consider the case that A and B are elementary abelian varieties.

Let q be a positive integer. Consider the exact sequence:

$$0 \longrightarrow \mathrm{Hom}(A,B) \xrightarrow{\times q} \mathrm{Hom}(A,B) \longrightarrow$$

$$\longrightarrow \mathrm{Hom}({}_qA,B) \longrightarrow \mathrm{Ext}(A,B) \xrightarrow{\times q} \mathrm{Ext}(A,B).$$

We study the following cases:

a) $\tau(A) \neq 0$ and $\tau(B) = 0$,
b) $\tau(A) = 0$ and $\tau(B) \neq 0$,
c) $\tau(A) \neq 0$ and $\tau(B) \neq 0$,
d) $\tau(A) = 0$ and $\tau(B) = 0$.

a&b) In these cases $\mathrm{Hom}(A,B) = 0$, and for $(q,p) = 1$, $\mathrm{Hom}({}_qA,B) \neq 0$ (or: $A \oplus B$ has only A and B as abelian subvarieties, consider $A \oplus B/N$, with $N \not\subset A$ and $N \not\subset B$, etc.).

c) Hom(A,B) is a finitely generated $\mathbb{Z}$-module, and $\mathrm{Hom}({}_qA,B)$, $q = p$, is a k-vectorspace not equal to zero; hence in this case

$$\mathrm{Hom}(A,B) \longrightarrow \mathrm{Hom}({}_pA,B)$$

is not surjective.

d) Let $\dim A = m$ and $\dim B = n$; ${}_pA = (\nu_p)^m \oplus (\mu_p)^m$, hence

$$\mathrm{Hom}({}_pA,B) = (\mathbb{Z}/p\mathbb{Z})^{2mn} .$$

Thus the rank of the (free) $\mathbb{Z}$-module Hom(A,B) is at most 2mn. Let q be a prime number different from p. As ${}_qA = (\mu_q)^{2m}$,

$$\mathrm{Hom}({}_qA,B) = (\mathbb{Z}/q\,\mathbb{Z})^{4mn} .$$

Thus

$$\mathrm{Hom}(A,B) \longrightarrow \mathrm{Hom}({}_qA,B)$$

is not surjective.

QED

<u>REMARK</u>: It is known that the free $\mathbb{Z}$-module Hom(A,B) has finite rank, and that its rank is at most $4n^2$ in case $A = B$, (cf. [41] , page 141). By the calculations just made, it is clear that this extreme value only can be realized in the case $\tau(A) \neq 0$. One can easily see in which cases the p-primary and the q-primary part of Hom(A,B) is zero. For example: the q-primary part, $(q,p) = 1$, in case (c) is zero if and only if the rank of Ext(A,B) equals 4mn.

<u>Alternative proof</u> of (14.1): Show that the injective dimension of an elementary object is at most two (e.g. μ_p : (8.7); $\underline{\underline{G}}_m$: (12.3); abelian variety: (12.4) and (12.5); etc.).

II.15 Some remarks and questions

(15.1) Colocalization - Let $\underline{A}$ be a locally artinian category (i.e. an abelian category with artinian cogenerators and exact filtered projective limits). Let Σ be the (multiplicative) set of positive integers. We denote by $\underline{A}_\Sigma$ the full subcategory of $\underline{A}$ defined by Σ: an object $M \in \underline{A}_\Sigma$ is the projective limit of its quotient objects N with $n.1_N$ for some $n \in \Sigma$ (dual situation: [17], III.5d). This is a colocalizing subcategory, i.e.

$$T : \underline{A} \longrightarrow \underline{A}/\underline{A}_\Sigma = \underline{D}$$

has a left adjoint $S : \underline{A}/\underline{A}_\Sigma \longrightarrow \underline{A}$, the cosection. It is not difficult to show that if

$$(*) \quad \mathrm{Hom}_{\underline{A}}(\varprojlim_J G_j, H) \xrightarrow{\sim} \varinjlim_J \mathrm{Hom}_{\underline{A}}(G_j, H)$$

is an isomorphism,

$$\mathrm{Ext}^n_{\underline{A}}(G,H) \otimes_{\mathbb{Z}} \mathbb{Q} \cong \mathrm{Ext}^n_{\underline{D}}(TG,TH).$$

This can be applied to the situation $\underline{A} = \underline{P}$, or to the category of projective systems of quasi-algebraic groups (the category denoted by $\underline{P}$ in GP will be indicated by $\underline{B}$ here). Clearly the obvious functor

$$U : \underline{P} \longrightarrow \underline{B}$$

induces an equivalence between $\underline{P}/\underline{P}_{loc}$ and $\underline{B}$. As $\underline{P}_{loc} \subset \underline{P}_\Sigma$, the categories $\underline{P}/\underline{P}_\Sigma$ and $\underline{B}/\underline{B}_\Sigma$ are equivalent. Hence:

$$\mathrm{Ext}^n_{\underline{P}}(G,H) \otimes_{\mathbb{Z}} \mathbb{Q} \cong \mathrm{Ext}^n_{\underline{B}}(UG,UH) \otimes_{\mathbb{Z}} \mathbb{Q}, \qquad G \in \underline{P},\ H \in \underline{G}.$$

Thus a (multiple) extension of group schemes defines a torsion element if and only if the corresponding extension of quasi-algebraic groups defines a torsion element.

(15.2) Homological dimension - Let $\underline{A}$ be a locally artinian category. Then the projective envelopes of objects of $\underline{A}_{\Sigma}$ belong to $\underline{A}_{\Sigma}$ (cf. [17], III.5, proposition 12). Suppose that for all H running through a set of cogenerators the condition (*) holds for all $G = \varprojlim G_j$. Then

$$\text{hd } \underline{A} = \max(\text{hd } \underline{A}_{\Sigma}, \text{hd } \underline{D}),$$

or

$$\text{hd } \underline{A} = \max(\text{hd } (\underline{A}_{\Sigma}) + 1, \text{hd } \underline{D}).$$

This fact is not very difficult to prove; one has to apply the dual of [17], page 376, corollary 5, the result mentioned in (15.1), and a little diagram chasing.

By GP we know that the homological dimension of $\underline{P}/\underline{P}_{\Sigma}$ is one. The objects $\underline{\underline{G}}_m$, A, where A is running through a set of elementary abelian varieties containing one copy of each isogeny type, form a set of cogenerators for $\underline{P}/\underline{P}_{\Sigma}$. As multiplication by an integer is an epimorphism for these objects, the result just mentioned yields:

$$\text{hd } \underline{P} = \max(\text{hd } \underline{P}_{\Sigma}, \text{hd } \underline{P}/\underline{P}_{\Sigma}).$$

Thus the shortest proof of (14.1) seems this method combined with the fact $\text{hd } \underline{P}_{\Sigma} = 2$. The proof of (14.1) we gave in the sections 7 - 12 consists of proving $\text{hd } \underline{P}_{\Sigma} = 2$, and following the proof of the properties just mentioned in the case $\underline{A} = \underline{P}$.

It seems desirable to have general principles to deduce hd $\underline{A}$ from hd $\underline{C}$ and hd $\underline{A}/\underline{C}$ in case $\underline{C}$ is a thick (or a localizing) subcategory of $\underline{A}$. Is it possible that $\text{hd } \underline{C} > \text{hd } \underline{A}$?

The result $\text{hd } \underline{P}_{\Sigma} = 2$ can also be deduced from a theory recently developed by CARTIER and GABRIEL (unpublished):

(15.3) Unipotent group schemes and A-modules - Let $\underline{W} = \underline{W}(k)$ be the ring of infinite Witt-vectors over k; an element $w \in \underline{W}$ is a sequence

$$w = (w_0, w_1, \ldots), \quad w_i \in k,$$

and addition and multiplication are given by certain polynomials;

$$p\cdot(w_0, w_1, \ldots) = (0, w_0^p, w_1^p, \ldots);$$

we write

$$(w_0, w_1, \ldots)^\sigma = (w_0^p, w_1^p, \ldots);$$

$\underline{W}$ is a complete local ring of characteristic zero with maximal ideal $p\underline{W}$, and residue class field $\underline{W}/p\underline{W} \cong k$; these properties of $\underline{W}$ determine it uniquely (k being perfect), and σ is the unique automorphism inducing the automorphism of raising to p-th powers in the residue class field (compare: J.-P.Serre - Corps locaux, II.5&6). We denote by A the (non-commutative) ring generated over $\underline{W}$ by F and V with the relations

$$Fw = w^\sigma F, \quad wV = V w^\sigma, \quad FV = VF = p$$

(in [17], page III-14, this ring was denoted by A"). Every element $x \in A$ can be written uniquely in the form

$$x = w + \sum^{<\infty} a_i F^i + \sum^{<\infty} b_j V^j, \qquad w, a_i, b_j \in \underline{W}.$$

It can be proved that

$$A_n \overset{\text{def}}{=\!=} A/(V^n) \cong \operatorname{Hom}(W_n, W_n)$$

is the endomorphismring of W_n (where (V^n) denotes the two-sided ideal $(V^n) = A\cdot V^n = V^n\cdot A$, k being perfect).

For an algebraic group scheme G we define

$$D(G) = \varinjlim_n \operatorname{Hom}(G, W_n).$$

In a natural way D(G) is a left A-module, and D is a functor. CARTIER and GABRIEL prove that D is an anti-equivalence between the category of finitely generated A-modules annihilated by some power of V, and the category of unipotent algebraic group schemes over k (k is supposed to be a perfect field in their theory). Examples:

$$D(\underline{\underline{G}}_a) \cong A/A\cdot V, \quad D(\alpha_p) \cong A/A\cdot(F,V), \quad D(\nu_p) \cong A/A\cdot(F-1,V).$$

The ring A is a subring of the ring E of non-commutative formal power series in F and V over $\underline{W}$ with the same relations explained above (cf. [23], page 21). We are going to use this result of CARTIER and GABRIEL, and methods of MANIN (cf. [23]), in order to make some calculations.

(15.4) Finite subgroups of abelian varieties - Let N be a finite group scheme over Spec(k). We claim that there exists an abelian variety which contains N as a subgroup scheme. If $N \in \underline{N}_{red}$ or $N \in \underline{N}_{lr}$ this is clear. Thus it suffices to prove the case $N \in \underline{N}_{ll}$. We settle this question here with the help of a theorem of MANIN: every commutative formal group is weakly algebroid (cf. [23], IV.4, theorem 4.2); in (15.11) we present another proof, based on an unpublished result of P.Russell.

As $L_{\infty,\infty}$ is the projective envelope of the only simple object of $\underline{N}_{ll}$, there exist numbers a,b,c and an epimorphism

$$(L_{a,b})^c \longrightarrow N^D.$$

By duality, there exists a monomorphism

$$N \hookrightarrow (L_{b,a})^c.$$

Thus it suffices to prove that $L_{b,a}$ can be embedded into an abelian variety. The formal group $G_{n,m}$ corresponds to the Dieudonné module

$E/E(F^m - V^n) = M_{n,m}$ (cf. [23], page 35), while $L_{n,m}$ corresponds to

$$D(L_{n,m}) \cong E/E(F^m, V^n) \cong A/A(F^m, V^n).$$

Thus $L_{n,m} \subset G_{n,m}$. Let X be an abelian variety. Then $\hat{X}$ (the completion of X along the zero point) is isogeneous to a formal group of type

$$\hat{X} \sim P = r_0 G_{1,0} + \Sigma\, r_{n,m} G_{n,m} \quad \begin{cases} 0 < n < \infty \\ 0 < m < \infty \\ (n,m) = 1. \end{cases}$$

Conversely, for every n and m with $(n,m) = 1$, there exists an abelian variety X such that $\hat{X}$ contains $G_{n,m}$ up to isogeny. Let M be the kernel of $\phi : P \longrightarrow \hat{X}$; then M is a finite formal group, hence M is a finite group scheme. Thus there exists an integer q such that $q \cdot 1_M = 0$. We write $M_1 = \mathrm{Ker}(P \xrightarrow{q \cdot 1_P} P)$, and $M_2 = \phi(M_1)$. As M_1 is a finite formal group, M_2 is a subgroup scheme of X; we define: $Y = X/M_2$. We thus obtain a commutative, exact diagram

$$\begin{array}{ccccc} & 0 & & 0 & \\ & \downarrow & & \downarrow & \\ M \subset & M_1 & \longrightarrow & M_2 & \longrightarrow 0. \\ & \downarrow & & \downarrow & \\ & P & \xrightarrow{\phi} & X & \\ & \downarrow & & \downarrow & \\ & P & \hookrightarrow & Y & \end{array}$$

Thus we have proved: if $\hat{X}$ contains a factor $G_{n,m}$ up to isogeny, then X is isogeneous to an abelian variety Y which contains $L_{n,m}$. For fixed n and m there exists such an abelian variety X, by the result of MANIN, and we have proved what we claimed.

(15.5) <u>Extensions of α_p over α_p</u> - Let N be a finite group scheme, containing α_p such that $N/\alpha_p \cong \alpha_p$. Which possibilities there

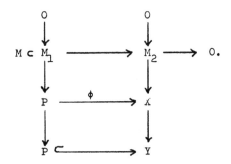

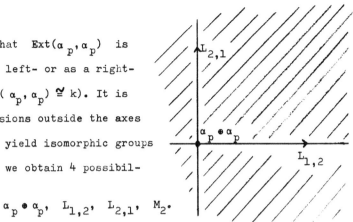

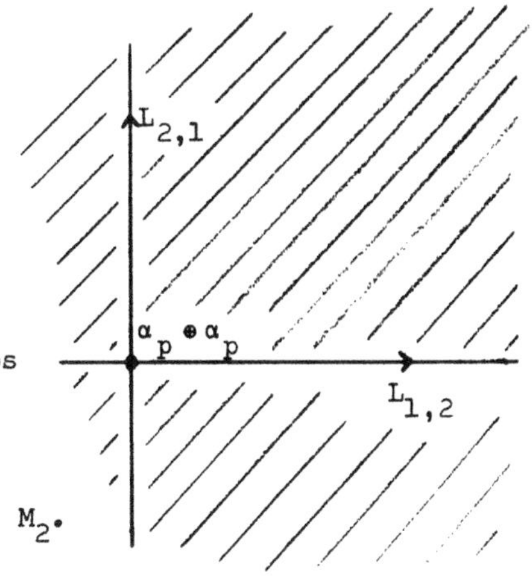

are for N? We know that $\text{Ext}(\alpha_p, \alpha_p)$ is of dimension 2 (as a left- or as a right-vectorspace over $\text{Hom}(\alpha_p, \alpha_p) \cong k$). It is clear that all extensions outside the axes $k \cdot L_{1,2}$ and $k \cdot L_{2,1}$ yield isomorphic groups (k is perfect); thus we obtain 4 possibilities:

$$\alpha_p \oplus \alpha_p, \quad L_{1,2}, \quad L_{2,1}, \quad M_2.$$

Another way of proving this, is described in (15.4): consider all A-modules of length 2; then we obtain:

$$D(\alpha_p \oplus \alpha_p) \cong k \oplus k, \quad D(L_{1,2}) \cong A/A(F^2, V),$$

$$D(L_{2,1}) \cong A/A(F, V^2), \quad D(M_2) \cong A/A(F^2, F-V, V^2).$$

This can be used as follows. Let C be an abelian variety of dimension one (i.e. an elliptic curve) with $\sigma(C) = 0$ (i.e. exceptional type; i.e. Hasse invariant equal to zero; i.e. no points of order p). Then ${}_pC = \underline{\underline{I}}^2(C) \cong M_2$; this can be seen as follows: the possibilities ${}_pC = \alpha_p \oplus \alpha_p$, ${}_pC = L_{2,1}$ are excluded as the dimension of C is one, and $({}_pC)^D = {}_p(C^t) = {}_pC$, hence ${}_pC = L_{1,2}$ is excluded.

(15.6) $\text{Ext}(\alpha_p, \alpha_p)$ <u>as a right- and as a left</u>-k-<u>vectorspace</u> - As $k \cong \text{Hom}(\alpha_p, \alpha_p)$ (ringisomorphism), we can consider $\text{Ext}(\alpha_p, \alpha_p)$ as a right- and as a left-k-vectorspace. We denote by $\xi_{1,2}$ the element of it defined by the exact sequence

$$0 \longrightarrow \alpha_p \xrightarrow{\ i\ } L_{1,2} \xrightarrow{\ F\ } \alpha_p \longrightarrow 0$$

(cf. page 10-2), and by $\xi_{2,1}$ we denote the element defined by the exact sequence

$$0 \longrightarrow \alpha_p \xrightarrow{\;V\;} L_{2,1} \xrightarrow{\;R\;} \alpha_p \longrightarrow 0.$$

Let $a \in \mathrm{Hom}(\alpha_p, \alpha_p)$; direct verification shows:

$$a_*(\xi_{1,2}) = a\cdot \xi_{1,2} = \xi_{1,2}\cdot a^p = (a^p)^*(\xi_{1,2})$$

and

$$a^*(\xi_{2,1}) = \xi_{2,1}\cdot a = a^p\cdot \xi_{2,1} = (a^p)_*(\xi_{2,1}).$$

(15.7) <u>In general</u> τ <u>is not invariant under isogeny</u> - We reproduce an example of BARSOTTI (cf. [2] , page 24). Let C be an abelian variety of dimension one with $\tau(C) = 1$. Let $i, j \in \mathrm{Hom}(\alpha_p, C)$ with $i \neq 0 \neq j$. We define A by the exact sequence

$$0 \longrightarrow \alpha_p \xrightarrow{\;(i,j)\;} C \times C \longrightarrow A \longrightarrow 0, \qquad \frac{i}{j} = a \in k.$$

If $a^{p^2} \neq a$, then $\tau(A) = 1$. This we prove as follows. Consider the diagram

$$\begin{array}{ccccccc}
0 \longrightarrow & \alpha_p & \longrightarrow & C \times C & \longrightarrow & A & \longrightarrow 0 \\
 & \Big\| & & p \downdownarrows q & & \Big\downarrow & \\
0 \longrightarrow & \alpha_p & \overset{i}{\underset{j}{\rightrightarrows}} & C & \longrightarrow & B & \longrightarrow 0
\end{array}$$

(p and q are the two projections), which is exact and commutative for the pair (p,i) and for (q,j). We obtain a diagram

$$\begin{array}{ccccccc}
 & \mathrm{Hom}(\alpha_p, C\times C) & \longrightarrow & \mathrm{Hom}(\alpha_p, A) & \longrightarrow & \mathrm{Ext}(\alpha_p, \alpha_p) & \xrightarrow{\;(i,j)_*\;} & \mathrm{Ext}(\alpha_p, C\times C) \\
\mathrm{Hom}(\alpha_p,\alpha_p) & p_* \downdownarrows q_* & & \Big\downarrow & & \Big\| & & p_* \downdownarrows q_* \\
\quad \llcorner\!\xrightarrow{\;\sim\;} & \mathrm{Hom}(\alpha_p, C) & \longrightarrow & \mathrm{Hom}(\alpha_p, B) & \overset{\partial}{\underset{\delta}{\rightrightarrows}} & \mathrm{Ext}(\alpha_p, \alpha_p) & \overset{i}{\underset{j}{\rightrightarrows}} & \mathrm{Ext}(\alpha_p, C)
\end{array}$$

Clearly δ and ∂ are injective (as $\dim C = 1$). Let $0 \neq \xi \in \mathrm{Im}(\delta)$. Then

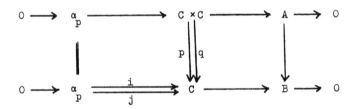

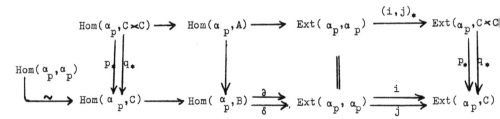

$$\xi = \beta\, \xi_{1,2} + \gamma\, \xi_{2,1}\,, \quad \text{and} \quad \beta \neq 0 \neq \gamma$$

as ${}_pC \cong M_2$ (cf. 15.5). Then $\mathrm{Im}(\delta) = \xi \cdot k$, and $\mathrm{Im}(\partial) = (a\cdot\xi)\cdot k$;

$$a\cdot\xi = \beta\, \xi_{1,2}\cdot a^p + \gamma\, \xi_{2,1} \cdot \sqrt[p]{a}$$

(cf. 15.6), thus $\mathrm{Im}(\delta) \cap \mathrm{Im}(\partial) = 0$. Thus $(i,j)_*$ is injective, and it follows that $\dim_k \mathrm{Hom}(\alpha_p, A) = 1$.

Another example is the following. Let X be an abelian variety with $\hat{X} = G_{1,2} \oplus G_{2,1}$; then $\underline{\underline{I}}^1(X) \cong \alpha_p \oplus L_{2,1}$. Define

$$(1,i): \alpha_p \longrightarrow \alpha_p \oplus L_{2,1} \subset X, \qquad Z = X/\mathrm{Im}(1,i).$$

It can be proved that $\tau(Z) = 1 \neq 2 = \tau(X)$. Further examples will be given in (15.11).

(15.8) <u>What is the maximum for</u> $\tau(-) + \sigma(-)$ <u>within isogeny type?</u>
Let X be an abelian variety. If $\dim(X) - \sigma(X) = 2$, there exists an abelian variety Z isogeneous to X such that $\sigma(Z) + \tau(Z) = \dim(Z)$; this was already known by BARSOTTI. We prove it as follows: the isosimple unipotent factors of $\hat{X}$ are isomorphic with $G_{1,1}$ by the symmetry condition of MANIN (cf. [23], IV.3, theorem 4.1; compare also page 19-3); choose Z such that $\hat{Z} \cong 2G_{1,1} + \ldots$ (compare 15.4). In that case $(\underline{\underline{I}}^1(Z))_{11} \cong \alpha_p \oplus \alpha_p$, hence $\tau(Z) = 2$.

Next we give an example of an abelian variety X of dimension 3 with $\sigma(X) = 0$ such that any Z isogeneous to X has $\tau(Z) = 1$ or $\tau(Z) = 2$ (hence $\tau(Z) + \sigma(Z) = 3$ is excluded in this case: this answers a question of BARSOTTI, cf. [2], page 24). We choose for X an abelian variety such that

$$\hat{X} \cong G_{1,2} \oplus G_{2,1}$$

(cf. [23], page 77/78, example 1; if necessary, replace X within isogeny type). Let $f: X \longrightarrow Z$ be an isogeny. We define N by:

$$\begin{array}{ccc} X \times_Z \underline{\underline{I}}^1(Z) = N & \longrightarrow\!\!\!\!\rightarrow & \underline{\underline{I}}^1(Z) \\ \downarrow & & \downarrow \\ X & \xrightarrow{f} & Z . \end{array}$$

$D(N)$ is a quotient of $M_{1,2} \oplus M_{2,1}$ and $D(\underline{\underline{I}}^1(Z)) \overset{\text{def}}{=\!=} U \subset D(N)$.

We choose integers n and m such that a factorization:

$$\begin{array}{ccc} M_{1,2} \oplus M_{2,1} & \longrightarrow & D(N) \\ \downarrow & \nearrow & \\ Q_1 \oplus Q_2 \overset{\text{def}}{=\!=} E/E(F-V^2,V^n) \oplus E/E(V-F^2,F^m) & & \end{array}$$

exists. We define U' by:

$$\begin{array}{ccc} Q_1 \oplus Q_2 & \hookleftarrow & U' \\ \downarrow\!\!\downarrow & & \downarrow\!\!\downarrow \\ D(N) & \hookleftarrow & U. \end{array}$$

Clearly $\tau(Z) = \tau(U) \leq \tau(U')$ (we write $\tau(-)$ instead of $\tau(D(-))$).

We define U_1' and U_2' by the exact, commutative diagram

$$\begin{array}{ccccccccc} & & 0 & & 0 & & 0 & & \\ & & \downarrow & & \downarrow & & \downarrow & & \\ 0 & \longrightarrow & U_1' & \longrightarrow & U' & \longrightarrow & U_2' & \longrightarrow & 0 \\ & & \downarrow & & \downarrow & & \downarrow & & \\ 0 & \longrightarrow & Q_1 & \longrightarrow & Q_1 \oplus Q_2 & \longrightarrow & Q_2 & \longrightarrow & 0 . \end{array}$$

An element $q \in Q$ can be written in the form

$$q = \alpha_i V^i + \ldots + \alpha_{n-1} V^{n-1} \quad \text{mod } E(F-V^2, V^n), \qquad \alpha_i \in k .$$

Thus it follows that any submodule of Q_1 is generated by one element. Hence $\tau(U_1') \leq 1$. In an analogous way we show that $\tau(U_2') \leq 1$, and we conclude $\tau(U') \leq 2$. Hence $\tau(Z) \leq 2$.

This indicates in which way perhaps an upper bound for $\tau(Z)$, Z isogeneous to X, can be found: let the unipotent part of $\hat{X}$ be isogeneous to $\Sigma_i\ G_{n_i,m_i}$; we conjecture: $\tau(Z) \leq \Sigma_i\ \min(n_i, m_i)$.

(15.9) Elementary abelian varieties of mixed type - Let $X \in \underline{G}$; we say that X is of mixed type if $(\underline{\underline{I}}^1 X)_{11} \neq 0$ and $(\underline{\underline{I}}^1 X)_{1r} \neq 0$.
Question: Does there exist an abelian variety of mixed type, which is elementary?

We could not settle this question. It seems that TATE constructed an example of an abelian variety of mixed type which is elementary (making use of zeta functions in the proof).

The only calculation in this direction we could make, is the following. Let $p = 3$, and let C be a hyperelliptic curve (i.e. a curve of genus 2), with Jacobian variety J. Then $\sigma(J) < 2$ implies that J is isogeneous to a product of two elliptic curves (non--isomorphic curves have non-isomorphic polarized Jacobian varieties, by Torelli's theorem, but their Jacobians may be isomorphic). We use notations of IGUSA, cf. [20], page 644/645: C can be of type n, with $1 \leq n \leq 6$. Suppose $n = 6$, i.e. C is given by

$$Y^2 = X(X-1)(X-\lambda_1)(X-\lambda_2)(X-\lambda_3) = \sum_{i=1}^{5} b_i X^i,$$

with $\lambda_1 = 1+\zeta$, $\lambda_2 = 1+\zeta+\zeta^2$, $\lambda_3 = 1+\zeta+\zeta^2+\zeta^3$, where ζ is a primitive fifth root of unity. We obtain (notation of [23], page 79):

$$A = \begin{pmatrix} b_2 & b_1 \\ b_5 & b_4 \end{pmatrix} \quad \text{(the Hasse-Witt matrix of J).}$$

Clearly $\det(A) \neq 0$, hence $n = 6$ implies $\sigma(X) = 2$. Thus $\sigma(J) < 2$ implies $n < 6$, and using [20], lemma 9 (or [20], page 648), we derive the desired result.

It seems that in case $p = 2$ the curve

$$Y^2 - Y = X^3 + \alpha X + \beta X^{-1}, \quad \beta \neq 0,$$

defines a Jacobian variety J, which could be a good candidate for an elementary abelian variety of mixed type: this curve has no automorphisms, except one of order two (cf. [20], page 645), and its Jacobian variety J is of mixed type (cf. [23], page 79/80); however we could not decide whether J is elementary or not.

(15.10) Extensions of algebraic groups (extracted from conversation and correspondence with P.Russell, University of California) - Let $\underline{G} = \underline{G}_k$ be the category of commutative algebraic group schemes over k. Consider the set $\underline{S}$ consisting of exact sequences

$$0 \longrightarrow F \longrightarrow G \longrightarrow H \longrightarrow 0,$$

with $F = F_{red}$ and $H = H_{red}$ (and hence $G = G_{red}$; i.e. "algebraic groups", and separable homomorphisms). The set $\underline{S}$ satisfies the conditions (P-1), ... , (P-4') of [44], XII (only the last axiom needs some verification: use the snake lemma); elements of $\underline{S}$ will be called separable short exact sequences. Let $X = X_{red}$, $Y = Y_{red}$; clearly

$$\mathrm{Ext}_{\underline{S}}(X,Y) \xrightarrow{\sim} \mathrm{Ext}_{\underline{G}}(X,Y)$$

is an isomorphism (and we write $\mathrm{Ext}(-,-)$). It can be proved that

$$\phi : E^2_{\underline{S}}(X,Y) \longrightarrow E^2_{\underline{G}}(X,Y)$$

is an isomorphism. First we remark that ϕ is injective for all reduced X and Y (this follows from the fact that the homomorphism on Ext is an isomorphism: same arguments as used in [33], 3.3). In the following cases ϕ is surjective too:

1) $X = L$, a connected, reduced, unipotent group scheme (i.e. successive extension of copies of $\underline{\underline{G}}_a$), and $Y = N \in \underline{N}_{rl}$. Consider an exact sequence $L_1/N = L_2$, with L_1 reduced, connected and unipotent, and use the fact that $E^2_{\underline{G}}(L,L_1) = 0 = E^2_{\underline{G}}(L,L_2)$ (plus the five lemma);

2) $X = A$, $Y = B$, both abelian varieties. Let $\xi \in E^2_{\underline{G}}(A,B)$; this is a torsion element; choose a positive integer n such that $n \cdot \xi = 0$. We write $N = ({}_nA)_{ll}$. As $\mathrm{Ext}(M,B) = 0$ in case $M \in \underline{N}_{lr}$ or $M \in \underline{N}_{red}$, we can write

$$\xi = \xi_1 \cdot \xi_2, \xi_1 \in \mathrm{Ext}(N,B), \qquad \xi_2 \in \mathrm{Ext}(A,N)$$

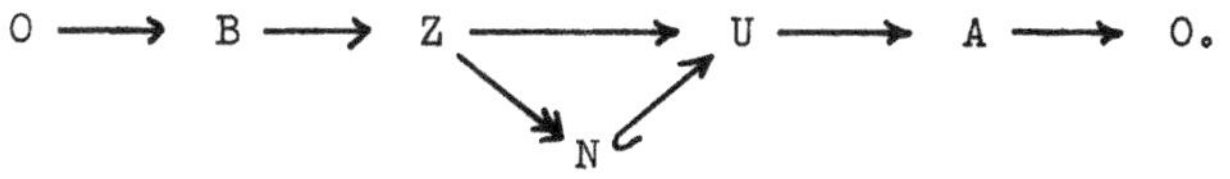

We can find a monomorphism $N \hookrightarrow (L_{b,a})^c$ (cf. 15.4), hence we can find a monomorphism $N \hookrightarrow (W_b)^c = G$ into a reduced, connected, unipotent algebraic group. By exactness of the sequence

$$\mathrm{Ext}(G,B) \longrightarrow \mathrm{Ext}(N,B) \longrightarrow E^2_{\underline{G}}(G/N,B) = 0$$

(cf. 12.8), we can write

$$\xi = \alpha_1 \cdot \alpha_2, \ \alpha_1 \in \mathrm{Ext}(G,B), \qquad \alpha_2 \in \mathrm{Ext}(A,G),$$

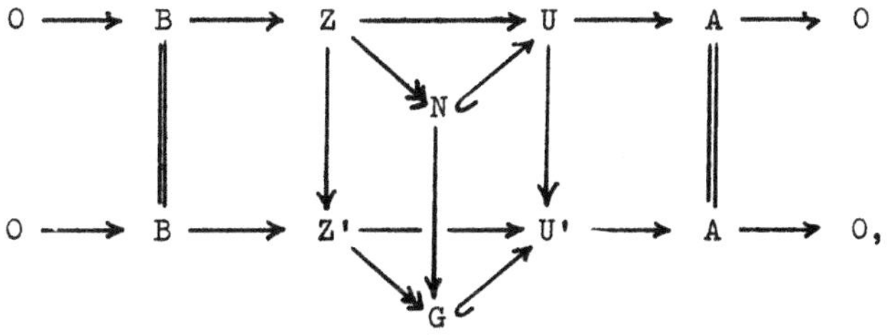

with Z' and U' reduced.

Now it follows easily that for reduced X and Y the map ϕ is an isomorphism. By (14.1) it now follows that $E^n_{\underline{S}}(X,Y)$ and $E^n_{\underline{G}}(X,Y)$ are naturally isomorphic for all n and for all reduced X and Y (k being an algebraically closed field).

We remark that P.RUSSELL has proved the following: let A and B be abelian varieties; then $E^2_{\underline{S}}(A,B) = 0$. We hope that the proof of this result will appear soon; once this result is established, the calculations summarized on page 14-2 are complete.

(15.11) <u>Finite subgroups of abelian varieties</u> (bis) - In this section we mention a consequence of the theorem of RUSSELL just mentioned. Let $N \in \underline{N}_{11}$, and let C be an abelian variety with $\tau(C) \neq 0$. Then N can be embedded into an abelian variety Z which is isogeneous to C^n, where n is some natural number. This we show by induction on the number of copies of α_p which occur if we write N as a successive extension of copies of α_p; this induction can start, as $\tau(C) \neq 0$ implies that there exists an embedding $\alpha_p \hookrightarrow C$. Suppose that $N_2/N_1 = N_3$, and suppose that $N_1 \subset B$, $N_3 \subset A$, where A and B are abelian varieties of the required type. Let $A' = A/N_1$. As $E^2(A',B) = 0$, an embedding of N_3 into an extension Y over A with kernel B results:

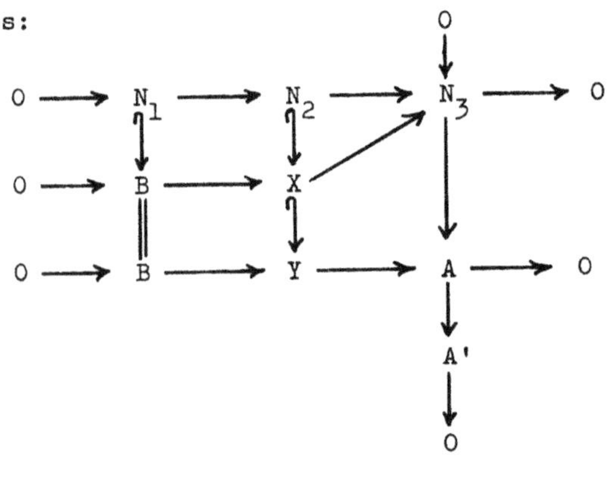

Let us consider the following particular case. We choose $N = L_{n,1}$

and for C we choose an elliptic curve of exceptional type. From the construction we performed it follows that we can find an embedding of $L_{n,1}$ into an abelian variety Z which is isogeneous to $X = C^n$. Hence in this case $\dim Z = n$, thus $\underline{\underline{I}}^1(Z) \cong L_{n,1}$. Thus for every positive integer n we see that C^n is isogeneous to an abelian variety Z with $\tau(Z) = 1$.

CHAPTER III DUALITY THEOREMS FOR ABELIAN SCHEMES

In this chapter we use the notations introduced in chapter I.

III.16 The CARTIER-SHATZ formula

In this section we prove a formula which can be found in papers by CARTIER (cf. [10] , also compare [11] , page 107, (22)); it was formulated by SHATZ in a form we are going to use it (cf. [38] , page 413, proposition 1).

[Gabriel: "Cette formule était connue de Serre, Grothendieck, Cartier depuis bien longtemps, en tout cas depuis 1959, sans que je sache exactement à qui elle est due."]

We prove the formula in the case of an arbitrary base scheme; the extra difficulties are only of technical character.

Let T be a prescheme, $X \in (Sch/T)$. We write:

$$\Gamma(X/T) = \mathrm{Mor}_T(T,X)$$

(notation of EGA, I.2.5.5; we use Mor to indicate a set of morphisms, and Hom_T to indicate a set of homomorphisms of group schemes over T). Let G be a group scheme over T. We write

$$H^T(G) = \mathrm{Hom}_T(G, \underline{\underline{G}}_{mT})$$

(the group of T-homomorphisms $G \longrightarrow \underline{\underline{G}}_{mT}$; for the definition of $\underline{\underline{G}}_{mT}$, compare section 1).

THEOREM (16.1) (CARTIER-SHATZ formula): Let T be a prescheme. Let N be a commutative group scheme over T, such that for every affine open set $U \subset T$, $N_U = \mathrm{Spec}(E_U)$ where E_U is a finitely generated, projective $\Gamma(U, \underline{O}_T)$-module (stronger condition: T is locally noetherian, and N is commutative, finite and flat over T). Then there exists an isomorphism

$$\Psi = \Psi(T,N) : H^T(N) \xrightarrow{\sim} \Gamma(N^D/T),$$

which is functorial in T and N.

REMARK: Let S be a locally noetherian prescheme, $T \in (Sch/S)$ and M a finite, flat, commutative group scheme over S. The theorem in this case reads:

$$\mathrm{Hom}_T(M \times_S T, \underline{\underline{G}}_{mT}) \xrightarrow{\sim} \mathrm{Mor}_T(T, (M_T)^D).$$

Using the fact that $(M^D)_T$ and $(M_T)^D$ are isomorphic (cf. 2.9), and using the isomorphism

$$\mathrm{Mor}_S(T, M^D) \xrightarrow{\sim} \mathrm{Mor}_T(T, M_T^D) = \Gamma(M_T^D/T)$$

(cf. EGA, I.3.3.14), this yields an isomorphism

$$\mathrm{Hom}_T(M_T, \underline{\underline{G}}_{mT}) \xrightarrow{\sim} \mathrm{Mor}_S(T, M^D);$$

in other terms: the functor

$$T \longmapsto \mathrm{Hom}_T(M \times_S T, \underline{\underline{G}}_{mS} \times_S T),$$

$T \in (Sch/S)$, is represented by M^D. Choosing S to be the spectrum of a field, we recognize the formulation of SHATZ.

Clearly if suffices to prove the theorem in case of an affine base scheme (both functors are local in T, and use the functorial character of Ψ).

We introduce the following notations:

> B is a ring, $T = \mathrm{Spec}(B)$,
> E is a B - bialgebra, which is a finitely generated, projective B - module,
> $N = \mathrm{Spec}(E)$ is the corresponding group scheme over T.

Every morphism $\alpha \in \Gamma(N^D/T)$ corresponds to a homomorphism $E^D \longrightarrow B$ of B-algebras. Thus the following inclusion is clear:

$$\Gamma(N^D/T) = \mathrm{RHom}_B(E^D, B) \subset \mathrm{Hom}_B(E^D, B) = E^{DD}$$

(RHom_B denotes the set of homomorphisms of B-algebras); an element $\alpha \in \Gamma(N^D/T)$ and the corresponding one in E^{DD} will be identified. In the same way we obtain:

$$H^T(N) \subset \mathrm{Mor}_T(N, \underline{\underline{G}}_{mT}) \cong E^* \subset E$$

(E^* is the multiplicative group of units of the ring E). We identify $f \in H^T(N)$ and the corresponding element $x \in E$. We are going to prove that there exists a commutative diagram

$$\begin{array}{ccc} (N^D/T) & \hookrightarrow & E^{DD} \\ \uparrow \Psi & & \cong \uparrow \kappa_E \\ H^T(N) & \hookrightarrow & E \end{array}$$

(for the definition of κ_E, see section 2; we remark that in this case it is an isomorphism, cf. 2.3), and we show that Ψ thus defined is an isomorphism. Once arrived at that point, the theorem is proved as the functorial character of Ψ is clear by the analogous properties of κ .

We define

$$\Lambda = \{ x \in E \mid \varepsilon_E(x) = 1, \quad s_E(x) = x \otimes x \} .$$

We are going to show that there exist isomorphisms

$$H^T(N) \cong \Lambda \cong \Gamma(N^D/T)$$

making commutative the diagram

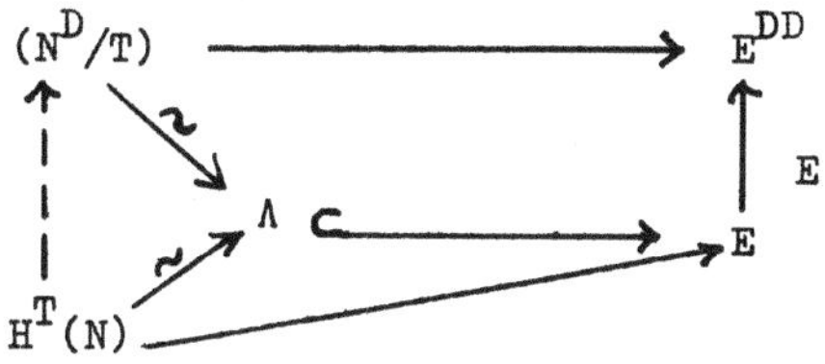

a) First we remark that $\Lambda \subset E^*$: the diagram

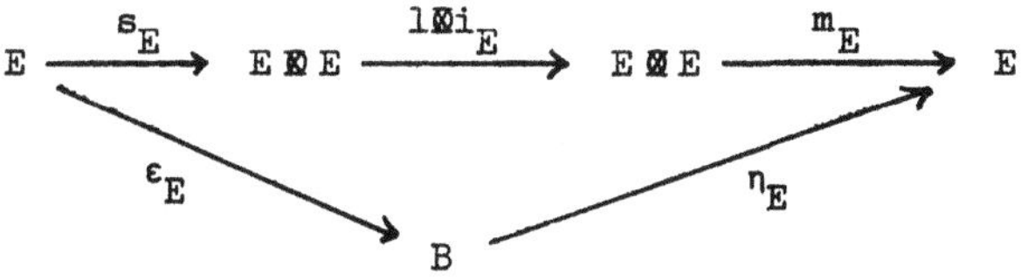

is commutative; hence $\varepsilon_E(x) = 1$ implies

$$x \cdot i_E(x) = \eta_E \, \varepsilon_E(x) = 1.$$

b) Suppose $x \in E^*$. Then $x \in \Lambda$ if and only if $x \in H^T(N)$. Consider the diagram

$\phi(X) = x$

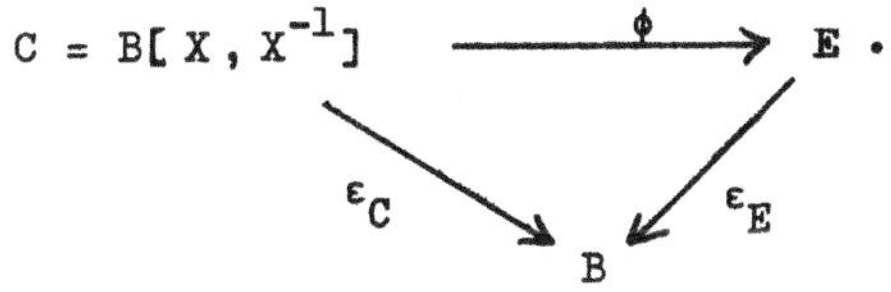

As $\varepsilon_C(x) = 1$, this diagram is commutative if and only if $\varepsilon_E(x) = 1$.

Consider the diagram

$$\begin{array}{ccc} E & \xleftarrow{\ \phi\ } & C \\ \big\downarrow{\scriptstyle s_E} & & \big\downarrow{\scriptstyle s_C} \\ E \otimes E & \xleftarrow{\ \phi \otimes \phi\ } & C \otimes C\,; \end{array}$$

This is commutative (i.e. $x \in H^T(N)$) if and only if

$$s_E(x) = s_E\ \phi(X) = \phi(X) \otimes \phi(X) = x \otimes x.$$

Hence $H^T(N) \cong \Lambda$.

c) Let $x \in E$, $\kappa_E(x) = f\colon E^D \longrightarrow B$, i.e.

$$(B \cong B^D \xrightarrow{\ f^D\ } E^{DD} \cong E) = g$$

with $g(1) = x$ and

$$(E^D \xrightarrow{\ g^D\ } B^D \cong B) = f.$$

Clearly $\varepsilon_E(x) = 1$ if and only if $f(1) = 1$. The diagram

$$\begin{array}{ccc} E & \xleftarrow{\ g\ } & B \\ \big\downarrow{\scriptstyle s_E} & & \big\|{\scriptstyle 1_B} \\ E \otimes E & \xleftarrow{\ g \otimes g\ } & B \otimes B \cong B \end{array}$$

is commutative if and only if the diagram

$$\begin{array}{ccc} E^D & \xrightarrow{\ g^D\ } & B^D \cong B \\ \big\uparrow & & \big\| \\ E^D \otimes E^D & \xrightarrow{\ g^D \otimes g^D\ } & B^D \otimes B^D \cong B \end{array}$$

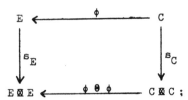

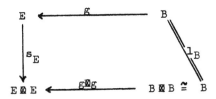

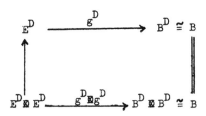

is commutative. Thus $x \in \Lambda$ if and only if $f: E^D \longrightarrow B$ is a ringhomomorphism, i.e. ${}^a f \in \Gamma(N^D/T)$. Thus $\Gamma(N^D/T) \cong \Lambda$ and the theorem is proved.

Let us consider an example. Choose a field of characteristic $p \neq 0$. Then:

<u>(16.2)</u>: $N \in \underline{N}_{ll}$ or $N \in \underline{N}_{rl}$ implies $\mathrm{Hom}(N, \underline{\underline{G}}_m) = 0$, and $N \in \underline{N}_{lr}$ or $N \in \underline{N}_{rr}$ and $N \neq 0$ imply $\mathrm{Hom}(N, \underline{\underline{G}}_m) \neq 0$.

Hence α_p and ν_p are not subgroups of $\underline{\underline{G}}_m$ and μ_q can be embedded in $\underline{\underline{G}}_m$ for all q (as we know already).

Next we describe an "additive analogon" of the CARTIER-SHATZ formula:

<u>PROPOSITION</u> (16.3): Let $T = \mathrm{Spec}(B)$, $N = \mathrm{Spec}(E)$ a commutative group scheme over T such that E is finitely generated and projective over B. There exists a (functorial) isomorphism

$$\mathrm{Hom}_T(N, \underline{\underline{G}}_{aT}) \xrightarrow{\sim} \mathrm{Der}_B(E^D).$$

[If F is a B-algebra with augmentation $\varepsilon_F: F \longrightarrow B$, we write

$$\mathrm{Der}_B(F) = \{ \phi \mid \phi \in \mathrm{Hom}_B(F,B), \text{ such that for } x,y \in F, \ \phi(xy) = \phi(x)\,\varepsilon(y) + \varepsilon(x)\,\phi(y)\} \quad .]$$

Sketch of the <u>proof</u>: Clearly $\mathrm{Hom}_T(N, \underline{\underline{G}}_{aT}) \subset E$ and $\mathrm{Der}_B(E^D) \subset E^{DD}$. An element $x \in E = \mathrm{Mor}_T(N, \underline{\underline{G}}_{aT})$ defines a T-morphism $x: N \longrightarrow \underline{\underline{G}}_{aT}$. This is a homomorphism if and only if

$$s_E(x) = x \otimes 1 + 1 \otimes x.$$

One easily proves this to be the case if and only if $\kappa_E(x) = \alpha \in E^{DD}$

has the property

$$\alpha(\phi\psi) = \alpha(\phi)\,\varepsilon_{E^D}(\psi) + \varepsilon_{E^D}(\phi)\,\alpha(\psi).$$

The proof is concluded analogously to the proof of (16.1).

COROLLARY (16.4): Let $B = k$ be a field. If $N \in \underline{N}_{lr}$ or $N \in \underline{N}_{rr}$, then $\mathrm{Hom}(N, \underline{\underline{G}}_a) = 0$. If $N \neq 0$ and $N \in \underline{N}_{ll}$ or $N \in \underline{N}_{rl}$, $\mathrm{Hom}(N, \underline{\underline{G}}_a) \neq 0$.

In the case of characteristic $p \neq 0$, the only simple object of $\underline{N}_{ll}$ is α_p (cf. [16] , III.1, lemma1, and [38] , III.2, proposition 11).

PROOF: If $N^D \in \underline{N}_{red}$, E^D is generated by idempotents, hence any derivation $\phi : E^D \longrightarrow k$ is zero. If $N^D \in \underline{N}_{loc}$ and $N^D \neq 0$, the k-vectorspace I/I^2 is different from zero, where $I = \mathrm{Ker}(\varepsilon_{E^D})$. Hence the k-linear dual of I/I^2, which is isomorphic with $\mathrm{Der}_k(E^D)$, is not zero.

Let $N \in \underline{N}_{ll}$ be a simple object. Then $\mathrm{Hom}(N, \underline{\underline{G}}_a) \neq 0$. Thus $N \subset \underline{\underline{G}}_a$. Then $N \cap \alpha_p \neq 0$ (as $N \in \underline{N}_{loc}$), thus $N \subset \alpha_p$. As $F \in A_1$ has no proper factors in the ring A_1, it follows $N = \alpha_p$.

QED

III.17 Extensions and sheaves

The category of commutative group schemes over a fixed base prescheme is in general not an abelian category. However we like to study groups of extensions in this category. The category of sheaves of commutative groups on the category (Sch/S), topologized in some way, is an abelian one; in this category the usual constructions can be carried out. This general principle (I suppose it stems from GROTHENDIECK) we are going to apply in this section. We follow the conventions exposed by DEMAZURE (cf. SGAD, IV; compare also M. ARTIN, [1] ; VERDIER, SGAA).

Let S be a locally noetherian prescheme. We write $\underline{C}_S$ for the category of all preschemes over S (not necessarily locally noetherian), $\hat{\underline{C}}_S$ for the category of presheaves of sets (= contravariant functors with values in $\underline{Ens}$) on $\underline{C}_S$, and $\underline{H}_S$ for the category of presheaves of commutative groups on $\underline{C}_S$; then $\underline{H}_S$ is the category of commutative group objects in $\hat{\underline{C}}_S$.

On $\underline{C}_S$ the (fpqc)-topology is defined (SGAD, IV.6.3: T_1): consider $X \in \underline{C}_S$, open subschemes $X_\alpha \subset X$ which cover X:

$$\bigcup X_\alpha = X$$

(N.B. the set of indices need not to be finite), and faithfully flat, quasi compact morphisms $Y_\alpha \longrightarrow X_\alpha$; thus we obtain a morphism

$$\coprod Y_\alpha \longrightarrow X.$$

Morphisms of this type generate the (fpqc)-topology. If $X \longrightarrow S$ is quasi compact, a morphism $Y \longrightarrow X$ which is covering in the (fpqc)-topology can be refined (i.e. $Z \longrightarrow Y \longrightarrow X$) to a faithfully flat, quasi compact morphism $Z \longrightarrow X$.

We denote by $\tilde{\underline{C}}_S$ the category of sheaves of sets on $\underline{C}_S$ with respect to the (fpqc)-topology, and we write $\underline{F}_S$ for the category of sheaves of commutative groups on this topologized category. Clearly $\underline{F}_S$ is the category of commutative group objects in $\tilde{\underline{C}}_S$. The following inclusions (horizontally: full embeddings) are clear:

$$\begin{array}{ccccc} \underline{C}_S & \subset & \tilde{\underline{C}}_S & \subset & \hat{\underline{C}}_S \\ \cup & & \cup & & \cup \\ \underline{G}_S & \subset & \underline{F}_S & \subset & \underline{H}_S; \end{array}$$

we identify an object of $\underline{C}_S$ with the corresponding sheaf and with the corresponding presheaf; same remark for group schemes. Important fact: $\underline{F}_S$ is an abelian category (cf. [1] , Chapter II, theorem 1.6).

Let $F,G,H \in \underline{G}_S$. We say that the sequence

$$(1) \qquad 0 \longrightarrow F \xrightarrow{\ f\ } G \xrightarrow{\ p\ } H \longrightarrow 0$$

is <u>exact</u>, f and p being homomorphisms, if $F = \mathrm{Ker}(p)$ (i.e. the sequence is left exact in $\underline{G}_S$ and in $\underline{C}_S$), and if

$$(2) \qquad F \times G \underset{p_2}{\overset{s_G \cdot (f, 1_G)}{\rightrightarrows}} G \xrightarrow{\ p\ } H$$

is exact in $\underline{C}_S$.

<u>REMARK</u>: We say that (1) is a right-exact sequence in $\underline{G}_S$, if

$$0 \longrightarrow \mathrm{Hom}(H,X) \xrightarrow{\ p^*\ } \mathrm{Hom}(G,X) \xrightarrow{\ f^*\ } \mathrm{Hom}(F,X)$$

is exact for all $X \in \underline{G}_S$. This last condition is not sufficient in general to prove the second condition needed for exactness of (1) (it may happen that the cokernel of f exists in $\underline{G}_S$, but that it is not equal to the cokernel of $(p_2, s_G \cdot (f, 1_G))$ in $\underline{C}_S$).

<u>PROPERTY</u> (17.1): Let $G, H \in \underline{G}_S$, and let $p: G \longrightarrow H$ be a faithfully flat homomorphism. We write $F = \mathrm{Ker}(p)$,

$$(1) \qquad 0 \longrightarrow F \longrightarrow G \xrightarrow{\ p\ } H \longrightarrow 0.$$

Then:

a) F is flat;

b) the sequence (1) is exact in $\underline{F}_S$;

c) the sequence (1) is exact.

<u>PROOF</u>: The square

$$\begin{array}{ccc} G & \xrightarrow{\ p\ } & H \\ \uparrow & & \uparrow{\scriptstyle 0} \\ F & \longrightarrow & S \end{array}$$

is cartesian (i.e. $F = G \times_H S$), hence (a) follows.

Next we remark that $G \longrightarrow S$ is quasi compact, and that G is locally noetherian (also: $H \longrightarrow S$ is separated), hence p is quasi compact (cf. EGA, I.6.6.4). As p is covering in the (fpqc)-topology, it is an epimorphism in $\tilde{\underline{C}}_S$ (cf. SGAD, IV.4.4.3), thus it is an epimorphism in $\underline{F}_S$, and (b) follows (as $\underline{F}_S$ is an abelian category).

The fact (c) results from SGA, 1961, VIII.5.2.

QED

<u>LEMMA</u> (17.2): Let $p: G \longrightarrow H$ be an epimorphism in $\underline{F}_S$. Then it is an epimorphism of sheaves (i.e. an epimorphism in $\underline{C}_S$).

<u>PROOF</u>: Let $(F \xrightarrow{\ f\ } G) = \mathrm{Ker}(p)$. We denote by $G \xrightarrow{\ q\ } K$ the cokernel of $(p_2, s_G \cdot (f, 1_G))$ in $\underline{C}_S$. This is a group object in $\underline{C}_S$ (cf. SGAD, IV.5.2.1), hence there exists a commutative diagram

in $\underline{F}_S$:

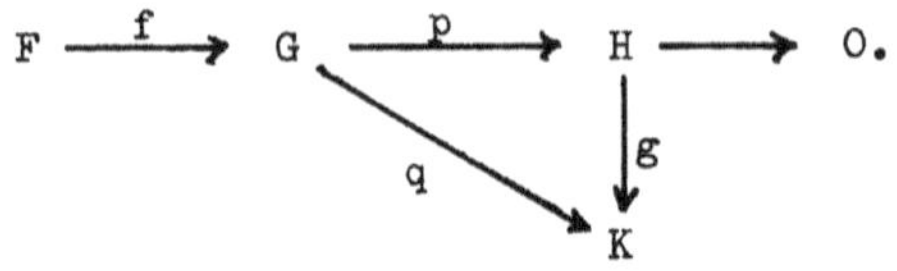

As $K = \mathrm{Coker}(p_2, s_G \cdot (f, 1_G))$ in $\underline{\tilde{C}}_S$, there exists a morphism of sheaves of sets $h: K \longrightarrow H$ such that

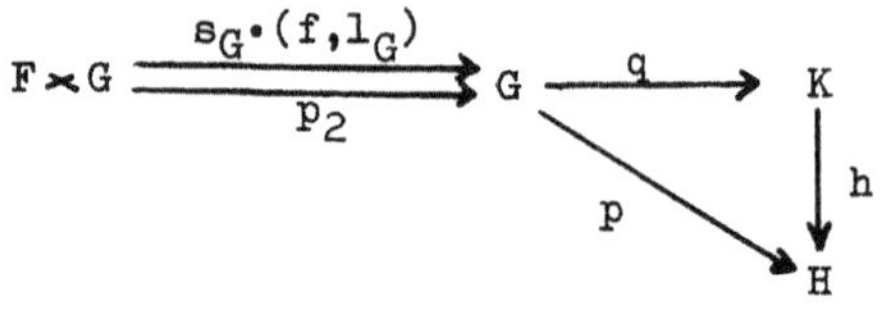

$hq = p.$

We show that h is a homomorphism (i.e. a morphism in $\underline{F}_S$). Consider the diagram in $\underline{C}_S$:

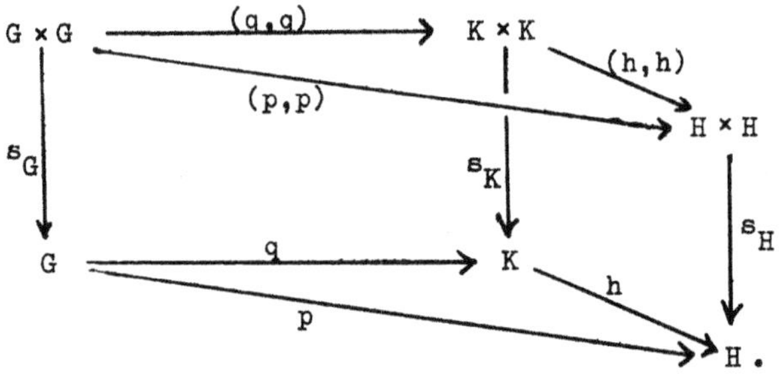

As q and p are homomorphisms,

$$s_H \cdot (h,h) \cdot (q,q) = s_H \cdot (p,p) = p \cdot s_G = h \cdot q \cdot s_G = h \cdot s_K \cdot (q,q).$$

As q is an epimorphism in $\underline{C}_S$, (q,q) is an epimorphism, hence

$$s_H \cdot (h,h) = h \cdot s_K;$$

thus h is a homomorphism of sheaves of groups. As $q = gp = ghq$, and as q is epimorphic (in $\underline{\tilde{C}}_S$ and in $\underline{F}_S$), $gh = 1_K$. As g and h are homomorphisms, and as p is epimorphic in $\underline{F}_S$, $p = hq = hgp$ implies

$hg = 1_H$. Thus $H = K$, and we have proved p to be an epimorphism in $\underline{\tilde{C}}_S$.

<u>PROPOSITION</u> (17.3): Let $G, H \in \underline{G}_S$, and let

$$p : G \longrightarrow H$$

be an epimorphism in $\underline{F}_S$. Suppose that p is a flat morphism (in that case $\mathrm{Ker}(p) = F$ is flat over S). Then p is faithfully flat (and the sequence

$$(1) \qquad 0 \longrightarrow F \xrightarrow{\ f\ } G \xrightarrow{\ g\ } H \longrightarrow 0$$

is exact, as follows by 17.1b).

<u>PROOF</u>: We write $F = \mathrm{Ker}(p)$; as $F = G \times_H S$ it follows that $F \in \underline{G}_S$ (i.e. F is representable). As p is epimorphic in $\underline{\tilde{C}}_S$ (cf. 17.2), this morphism is covering (cf. SGAD, IV.4.4.3). As H is quasi compact over S, there exists $X \in \underline{C}_S$ and a commutative diagram in $\underline{C}_S$:

$$\begin{array}{ccc} G & \xrightarrow{\ p\ } & H \\ & {\scriptstyle y}\nwarrow & \uparrow{\scriptstyle x} \\ & & X, \end{array}$$

with x faithfully flat and quasi compact. Thus $Y = G \times_H X$ is a principal homogeneous fibre space over X with structure group F, which is trivial. Thus we obtain a commutative diagram

$$\begin{array}{ccc} G & \xrightarrow{\ p\ } & H \\ \uparrow{\scriptstyle z} & & \uparrow{\scriptstyle x} \\ F \times X \cong Y & \xrightarrow{\ q\ } & X\ ; \end{array}$$

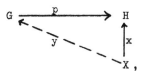

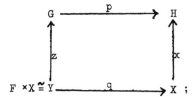

as F is faithfully flat and quasi compact over S, q is faithfully flat and quasi compact. Moreover z is faithfully flat (as it is obtained by base extension from x, cf. EGA, IV^2.2.2.13i). From the fact that xq and z are faithfully flat it follows that p is faithfully flat (cf. EGA, IV^2.2.2.13iii), and the proposition is proved.

PROPOSITION (17.4): Let

$$0 \longrightarrow F \xrightarrow{f} G \xrightarrow{p} H \longrightarrow 0$$

be an exact sequence in $\underline{F}_S$. Suppose that $F, H \in \underline{G}_S$, and that $F \longrightarrow S$ is an affine morphism. Then G is representable (i.e. $G \in \underline{G}_S$).

PROOF: As p is an epimorphism in $\underline{F}_S$, there exist $X \in \underline{C}_S$ and a commutative diagram

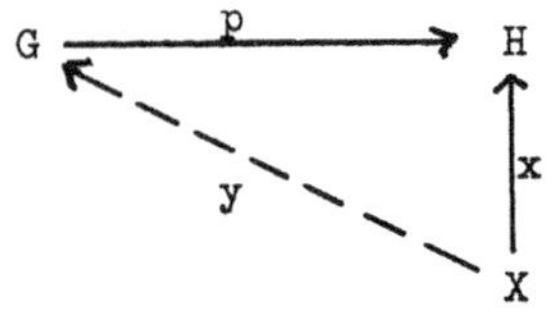

with x faithfully flat and quasi compact. Thus $Y = G \times_H X$ is a principal homogeneous fibre space over X with structure group F, which is trivial. Hence Y is representable (as F and H are representable). Moreover $q: Y \longrightarrow X$ is affine (as F is affine over S), hence the descent data given for Y (coming from G by base change $X \longrightarrow H$) are effective (cf. FGA, page 190-19, theorem 2), thus G is representable, and the proposition is proved.

DEFINITION: Let $F, H \in \underline{G}_S$, with $F \longrightarrow S$ flat. Consider an exact sequence

$$0 \longrightarrow F \xrightarrow{f} G \xrightarrow{p} H \longrightarrow 0$$

with $G \in \underline{G}_S$ and p faithfully flat and quasi compact. Two such sequences are called <u>equivalent</u>, if there exists a commutative diagram

$$\begin{array}{ccccccccc} & & & & G & & & & \\ & & & \overset{f}{\nearrow} & & \overset{p}{\searrow} & & & \\ 0 & \longrightarrow & F & & \downarrow g & & H & \longrightarrow & 0. \\ & & & \underset{f'}{\searrow} & & \underset{p'}{\nearrow} & & & \\ & & & & G' & & & & \end{array}$$

The set of equivalence classes is denoted by $\mathrm{Ext}_S(H,F)$ (remark that we defined in fact an equivalence relation: the sequences are exact in $\underline{F}_S$ by (17.1), hence $g\colon G \xrightarrow{\sim} G'$ is an isomorphism, in $\underline{G}_S$ or in $\underline{F}_S$).

<u>COROLLARY</u> (17.5): Let $H,F \in \underline{G}_S$ with F flat and affine over S. Then the map

$$\mathrm{Ext}_S(H,F) \xrightarrow{\sim} \mathrm{Ext}_{\underline{F}_S}(H,F)$$

is bijective (and hence $\mathrm{Ext}_S(H,F)$ is equiped with the structure of a commutative group). If

$$0 \longrightarrow H_1 \longrightarrow H_2 \longrightarrow H_3 \longrightarrow 0$$

is an exact sequence in $\underline{F}_S$ with $H_1,H_2,H_3 \in \underline{G}_S$, and $F \in \underline{G}_S$ is flat and affine over S, the usual exact $\mathrm{Hom}_S - \mathrm{Ext}_S$ sequence results.

<u>PROOF</u>: By (17.1) an exact sequence

$$0 \longrightarrow F \longrightarrow G \xrightarrow{p} H \longrightarrow 0$$

with p faithfully flat defines an element of $\mathrm{Ext}_{\underline{F}_S}(H,F)$. Hence $\mathrm{Ext}_S(H,F)$ can be considered as a subset of $\mathrm{Ext}_{\underline{F}_S}(H,F)$. By (17.4) and (17.3) a sequence exact in $\underline{F}_S$ defining an element of

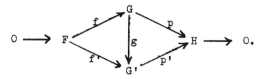

$\text{Ext}_{\underline{F}_S}(H,F)$ has a representable middle term, and it is an exact sequence. Thus the first half of the corollary is proved. The second statement then follows immediately.

REMARK: In case S is an artinian scheme, the hypothesis "F is affine" can be omitted, and the same result holds.

The notion of a primitive cohomology class will be taken from GA, VII.14. With this terminology we can obtain:

CONSTRUCTION and PROPOSITION (17.6): Let X be an abelian scheme over There exists an injective homomorphism

$$\alpha = \alpha_{X/S} : \text{Ext}_S(X, \underline{\underline{G}}_{mS}) \xrightarrow{\sim} H^1(X, \underline{O}_X^*),$$

which is functorial in X and in S, such that the image of α is the subgroup of primitive elements of $H^1(X, \underline{O}_X)$.

PROOF: An element $\xi \in \text{Ext}_S(X, \underline{\underline{G}}_{mS})$ is defined by an exact sequence

$$0 \longrightarrow \underline{\underline{G}}_{mS} \longrightarrow Y \xrightarrow{p} X \longrightarrow 0$$

with p faithfully flat and quasi compact. Hence Y is a principal homogeneous fibre space over X with structure group $\underline{\underline{G}}_{mS}$. Thus it is a locally trivial fibre space (cf. FGA, page 190-28, proposition 6.1; in the terminology of [35]: $\underline{\underline{G}}_{mS}$ is "special"). Hence this extension defines an element

$$\alpha(\xi) \in H^1(X, \underline{\text{Mor}}(X, \underline{\underline{G}}_{mS})) = H^1(X, \underline{O}_X^*)$$

(where $\underline{\text{Mor}}(-,-)$ denotes the sheaf of germs of morphisms). It is clear that $\alpha = \alpha_{X/S}$ is functorial in X and in S, and it follows that α is a homomorphism.

Suppose $\alpha(\xi) = 0$, i.e. the fibre space $Y \longrightarrow X$ to be trivial. As $\mathrm{Mor}_S(X,Y)$ is a group, we can choose a section $Y \xleftarrow{q} X$ in such a way that

$$\begin{array}{ccc} Y & \xleftarrow{\quad q \quad} & X \\ & {}_0\nwarrow \quad \nearrow_0 & \\ & S & \end{array}$$

commutes; we show that q is a homomorphism. Consider

$$t = (s_Y \cdot (q,q) - q \cdot s_X) : \ X \times X \longrightarrow Y.$$

Clearly $p \cdot t = 0$; thus t can be factored:

$$\begin{array}{ccc} & & X \times X \\ & {}^{u}\swarrow & \downarrow t \\ \underline{\underline{G}}_{mS} & \xrightarrow{\quad i \quad} & Y . \end{array}$$

We know that $(S \xrightarrow{0} X \times X \xrightarrow{u} \underline{\underline{G}}_{mS}) = (S \xrightarrow{0} \underline{\underline{G}}_{mS})$. By (5.1) we conclude:

$$\mathrm{Mor}_S(X \times X, \underline{\underline{G}}_{mS}) = \Gamma(S, \underline{O}_S)^* ,$$

and it follows that $t = 0$ ("every morphism of an abelian scheme into $\underline{\underline{G}}_{mS}$ is constant"). Thus t is a homomorphism, i.e. $\xi = 0$. This proves α to be injective.

From the functorial character of α it follows that the image of α contains only primitive cohomology classes.

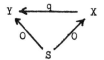

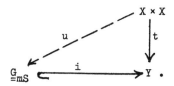

An element $c \in H^1(X, \underline{\mathrm{Mor}}(X, \underline{\underline{G}}_{mS})$ defines a principal homogeneous fibre space $p: Y \longrightarrow X$ with structure group $\underline{\underline{G}}_{mS}$. If c is primitive, there exists a morphism $s: G \times G \longrightarrow G$ such that the diagram

$$\begin{array}{ccc} Y \times Y & \xrightarrow{(p,p)} & X \times X \\ \downarrow{\scriptstyle s} & & \downarrow{\scriptstyle s_X} \\ Y & \xrightarrow{p} & X \end{array}$$

commutes. By standard techniques (cf. GA, page 183) we can equip Y with the structure of a group scheme with addition-morphism s, and an exact sequence

$$0 \longrightarrow \underline{\underline{G}}_{mS} \longrightarrow G \longrightarrow X \longrightarrow 0$$

results, which defines ξ such that $\alpha(\xi) = c$.

QED

<u>REMARK</u>: In the same way (plus Künneth formula) one establishes a natural isomorphism

$$\mathrm{Ext}_S(X, \underline{\underline{G}}_{aS}) \xrightarrow{\sim} H^1(X, \underline{O}_X)$$

for every abelian scheme X/S.

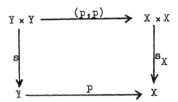

III.18 The generalized WEIL-BARSOTTI formula

In this section S denotes a locally noetherian prescheme, and X is an abelian scheme over S. We assume that either S is an artinian scheme or that X is projective over S.

THEOREM (18.1) (generalized WEIL-BARSOTTI formula): There exists a functorial isomorphism $\beta = \beta_{X/S}$ which is defined by the commutative diagram

$$\begin{array}{ccc} E^S(X) & \overset{\alpha}{\hookrightarrow} & Pic(X) \\ & \nwarrow{\scriptstyle \beta\,\sim} & \uparrow{\scriptstyle h} \\ & & \Gamma(X^t/S) \end{array} .$$

NOTATION: h will be defined below,

$E^S(X) = Ext_S(X, \underline{\underline{G}}_{mS}),$

$\Gamma(X^t/S) = Mor_S(S, X^t),$

$Pic(X) = H^1(X, \underline{O}_X^*).$

REMARK: Let k be a field, $S = Spec(k)$, and let X be an abelian variety over k. In that case the result is known:

$$Ext_k(X, \underline{\underline{G}}_m) \cong X_k^t \quad \text{(WEIL-BARSOTTI formula)}$$

(cf. GA, VII.16, theorem 6; for references, compare GA, page 196); this was proved by WEIL and by BARSOTTI. We remark that the proof of (18.1) we give below makes use of the (usual) formula.

PROPERTY (18.2): Let X,Y be abelian schemes over S, and let f: $X \longrightarrow Y$

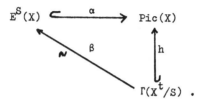

be a homomorphism such that for every connected component of S there exists a point s on it with $f(X_s) = 0 \in Y_s$. Then $f = 0$.

PROOF: By [27], 6.1, proposition 6.1, we know that there exists $\eta \in \Gamma(Y/S)$ such that

$$(X \xrightarrow{p} S \xrightarrow{\eta} Y) = f.$$

Thus

$$\eta = (S \xrightarrow{0} X \xrightarrow{p} S \xrightarrow{\eta} Y) = 0\,f = 0$$

(as f is a homomorphism), thus $f = 0\,p = 0$.

QED

LEMMA (18.3): Let X,Y be abelian schemes over S, and $f, g \in \mathrm{Hom}_S(X,Y)$. Suppose that Y^t exists. Then

$$(f+g)^t = f^t + g^t : Y^t \longrightarrow X^t.$$

PROOF (compare [27], page 118): Let

$$\rho = (f+g)^t - f^t - g^t.$$

For any $T \longrightarrow S$ we have $(f^t)_T = (f_T)^t$. Hence for $s = \mathrm{Spec}(k) \hookrightarrow S$ (where k is some field),

$$\rho_s = (f_s + g_s)^t - f_s^t - g_s^t.$$

By the theory of abelian varieties, it follows that $\rho_s = 0$ for all $s \in S$ (cf. [22], page 125, t2). Thus by (18.2) it follows $\rho = 0$, and the lemma is proved.

Let $T \in (Sch/S)$. The sequence

$$0 \longrightarrow \mathrm{Pic}(T) \longrightarrow \mathrm{Pic}(X \times T) \longrightarrow \underline{P}_{X/S}(T) \longrightarrow 0$$

is exact (this is the definition of $\underline{P}_{X/S}$, compare section 5). With the help of the morphism

$$(0,1_T) : T \longrightarrow X \times T$$

it is split exact. Thus we obtain a homomorphism

$$\sigma = \sigma_{X,T/S} : \underline{P}_{X/S}(T) \longrightarrow \mathrm{Pic}(X \times_S T),$$

which is functorial in $X,T/S$. Hence we can define an injective homomorphism

$$h_X = (\Gamma(X^t/S) \subset (\underline{\mathrm{Pic}}(X/S)/S) \cong \underline{P}_{X/S}(S) \overset{\sigma}{\hookrightarrow} \mathrm{Pic}(X)).$$

Using (17.6), the generalized WEIL-BARSOTTI formula follows from:

PROPOSITION (18.4): An element $\xi \in \mathrm{Pic}(X)$ is primitive if and only if $\xi \in \mathrm{Im}(h_X)$.

PROOF: Consider the homomorphisms

$$s_X , p_1 , p_2 : X \times X \longrightarrow X$$

(addition, respectively both projections). We define

$$\rho = s_X^* - p_1^* - p_2^* : \underline{\mathrm{Pic}}(X/S) \longrightarrow \underline{\mathrm{Pic}}(X \times X/S).$$

We are going to show that $\mathrm{Ker}(\rho) = X^t \subset \underline{\mathrm{Pic}}(X/S)$. By (18.3) we know that $\mathrm{Ker}(\rho) \supset X^t$. By the usual WEIL-BARSOTTI formula,

$$(\mathrm{Ker}(\rho))_s = \mathrm{Ker}(\rho_s) = (X^t)_s = X^t_s$$

for every $s \in S$. As X^t is an open subscheme of $\underline{\mathrm{Pic}}(X/S)$, this implies $\mathrm{Ker}(\rho) = X^t$.

By the functorial character of σ, it follows from $\mathrm{Ker}(\rho) \supset X^t$ that all elements contained in $\mathrm{Im}(h_X)$ are primitive.

Let $\xi \in \mathrm{Pic}(X)$ be a primitive element, and let $(0,1_S)^*(\xi) =$

$= \zeta \in \mathrm{Pic}(S)$. Clearly $\mathrm{Pic}(S \times_S S) = \mathrm{Pic}(S)$, and by the fact that ξ is primitive, it follows that $\zeta = \zeta + \zeta$. Thus $\zeta = 0$, i.e. $\xi \in \mathrm{Im}(\sigma: \underline{P}_{X/S}(S) \longrightarrow \mathrm{Pic}(X))$. Let $\xi = \sigma(g)$, where $g \in \underline{P}_{X/S}(S)$ corresponds with $g \in \Gamma(\underline{\mathrm{Pic}}(X/S)/S)$. As ξ is primitive,

$$(S \xrightarrow{g} \underline{\mathrm{Pic}}(X/S) \xrightarrow{\rho} \underline{\mathrm{Pic}}(X \times X/S)) = 0,$$

thus by $\mathrm{Ker}(\rho) \subset X^t$ this implies

$$g \in \Gamma(X^t/S) \subset (\underline{\mathrm{Pic}}(X/S)/S).$$

Thus $\xi \in \mathrm{Im}(h_X)$, and the proposition is proved; hence the proof of (18.1) is concluded.

The projections

$$p_1, p_2 : X \times_S X \longrightarrow X$$

and the injections

$$i_1 = (1,0), \ i_2 = (0,1) : X \longrightarrow X \times_S X$$

induce

$$p_1^t + p_2^t : X^t \times X^t \longrightarrow (X \times X)^t$$

and

$$(i_1^t, i_2^t) : (X \times X)^t \longrightarrow X^t \times X^t$$

(all products taken over S).

<u>COROLLARY</u> (18.5): These homomorphisms are inverse to each other (and hence $(X \times X)^t$ and $X^t \times X^t$ are isomorphic).

<u>PROOF</u>: Using (18.1), for any $T \in (Sch/S)$ we have

$$\mathrm{Mor}_S(T, (X \times X)^t) \cong E^T((X \times X)_T) \cong E^T(X_T) \times E^T(X_T) = (*)$$

and

$$\mathrm{Mor}_S(T, X^t \times X^t) \cong \mathrm{Mor}_S(T, X^t) \times \mathrm{Mor}_S(T, X^t) \cong E^T(X_T) \times E^T(X_T) = (**).$$

Clearly the identity between ($*$) and ($**$) is induced by the morphisms $p_1^t + p_2^t$ and (i_1^t, i_2^t).

QED

REMARK: Does it hold that "the connected component of $\underline{\text{Pic}}(Y \times Z/S)$ equals the product of the connected components of $\underline{\text{Pic}}(Y/S)$ and of $\underline{\text{Pic}}(Z/S)$" for arbitrary schemes $Y, Z/S$ (if $\underline{\text{Pic}}$.. exists, etc.)? For Picard varieties this is known (cf. [14] , page 486, theorem 2). Perhaps combination of this with [29] , page 294-14, lemma, yields more information.

Corollary (18.5) states that t is an additive functor. This can be reformulated:

THEOREM (18.6): The following diagrams are commutative:

a)

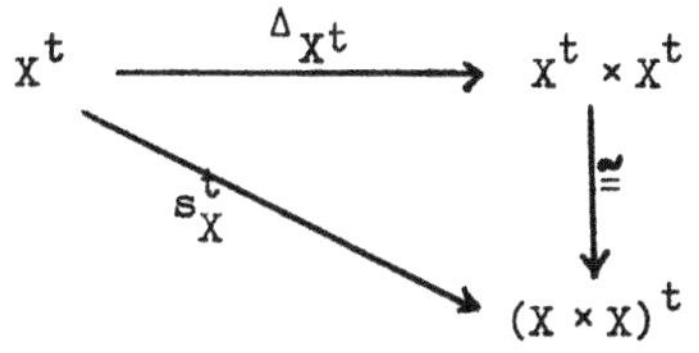

b)

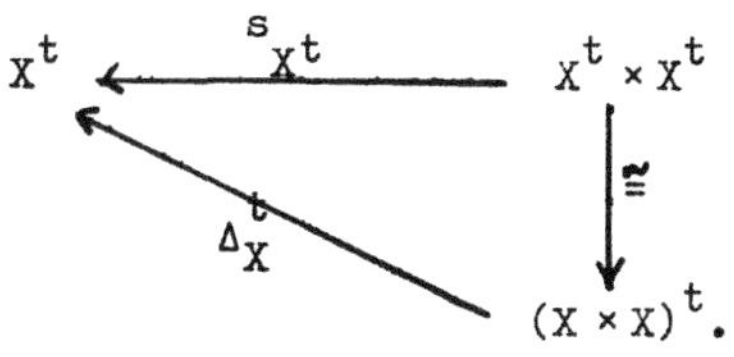

c) $\quad i_X^t = i_{X^t} \; : \; X^t \longrightarrow X^t.$

PROOF: a) and b) follow directly from (18.1). Using (18.3), the last statement follows: $0 = (1_X + i_X)^t = 1_X^t + i_X^t = 1_{X^t} + i_X^t$, and the theorem is proved.

REMARKS: Compare (18.6a) with [22] , VI.1, theorem 2.

Note the analogy in the formulas concerning finite group schemes and their duals (section 2), and abelian schemes and their duals.

III.19 The duality theorem

In this section S denotes a locally noetherian prescheme, and X and Y are abelian schemes over S. We assume either that S is an artinian scheme, or that X and Y are projective over S.

THEOREM (19.1): Let $\phi : X \longrightarrow Y$ be an isogeny (i.e. ϕ is surjective, hence ϕ is flat, and $\text{Ker}(\phi) = N$ is finite and flat over S, compare section 5, thus the sequence

$$(1) \qquad 0 \longrightarrow N \longrightarrow X \xrightarrow{\ \phi\ } Y \longrightarrow 0$$

is exact , cf. 17.1). Then there exists an exact sequence

$$(2) \qquad 0 \longrightarrow N^D \longrightarrow Y^t \xrightarrow{\ \phi^t\ } X^t \longrightarrow 0;$$

the construction of this sequence depends functorially on ϕ (i.e. if ϕ and ψ are isogenies between abelian schemes fitting into a commutative diagram

$$\begin{array}{ccc} X_1 & \xrightarrow{\ \phi\ } & Y_1 \\ \downarrow & & \downarrow \\ X_2 & \xrightarrow{\ \psi\ } & Y_2 \end{array},$$

we obtain a commutative diagram with exact rows:

$$\begin{array}{ccccccccc} 0 & \longrightarrow & (\text{Ker }\phi)^D & \longrightarrow & Y_1^t & \xrightarrow{\ \phi^t\ } & X_1^t & \longrightarrow & 0 \\ & & \uparrow & & \uparrow & & \uparrow & & \\ 0 & \longrightarrow & (\text{Ker }\psi)^D & \longrightarrow & Y_2^t & \xrightarrow{\ \psi^t\ } & X_2^t & \longrightarrow & 0 \end{array}).$$

PROOF: We write $M = \text{Ker}(\phi^t)$. Let $T \in (Sch/S)$. The sequence

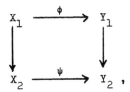

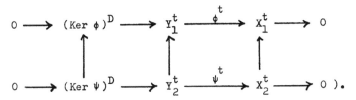

$$0 \longrightarrow M_T \longrightarrow Y_T^t \xrightarrow{\phi_T^t} X_T^t$$

is exact (distributivity of the fibred product), and the sequence

$$0 \longrightarrow N_T \longrightarrow X_T \xrightarrow{\phi_T} Y_T \longrightarrow 0$$

is exact (same argument, and 17.1). By (5.1) we conclude that

$$(\phi_T^t)^* : H^T(Y_T) \xrightarrow{\sim} H^T(X_T)$$

is an isomorphism (both isomorphic with $\Gamma(S,\underline{O}_S)^*$); hence by (17.5) we obtain an exact sequence

$$0 \longrightarrow H^T(N_T) \xrightarrow{\delta} E^T(Y_T) \xrightarrow{\phi_T^*} E^T(X_T)$$

(notation: $H^T(-) = \mathrm{Hom}_T(-,\underline{\underline{G}}_{mT})$, $E^T(-) = \mathrm{Ext}_T(-,\underline{\underline{G}}_{mT})$). By the generalized WEIL-BARSOTTI formula (cf. 18.1) we obtain an exact, commutative diagram

$$\begin{array}{ccccccc}
0 \longrightarrow & H^T(N_T) & \xrightarrow{\delta} & E^T(Y_T) & \xrightarrow{\phi_T^*} & E^T(X_T) \\
& \uparrow \cong \delta_{\phi_T} & & \uparrow \cong \beta_{Y_T} & & \uparrow \cong \beta_{X_T} \\
0 \longrightarrow & \Gamma(M_T/T) & \longrightarrow & \Gamma(Y_T^t/T) & \longrightarrow & \Gamma(X_T^t/T),
\end{array}$$

which defines an isomorphism

$$\delta = \delta_{\phi_T} : \Gamma(M_T/T) \xrightarrow{\sim} H^T(N_T)$$

(we identified $(X^t)_T$ and $(X_T)^t$). Using the CARTIER-SHATZ formula (cf. 16.1), we obtain an isomorphism

$$\begin{array}{ccccc} \Gamma(M_T/T) & \xrightarrow{\delta_{\phi_T}} & H^T(N_T) & \xrightarrow{\Psi} & \Gamma(N_T^D/T) \\ \| \wr & & & & \| \\ \mathrm{Mor}_S(T,M) & & \xrightarrow{\sim} & & \mathrm{Mor}_S(T,N^D) \end{array}$$

(we identify $(N^D)_T$ and $(N_T)^D$, N being finite and flat over T). Thus an isomorphism

$$\lambda_\phi : N^D = (\mathrm{Ker}\,\phi)^D \xrightarrow{\sim} M = \mathrm{Ker}(\phi^t)$$

results. Thus for each $s \in S$, $\phi_s^t : Y_s^t \longrightarrow X_s^t$ is surjective, hence ϕ^t is surjective, hence (cf. [27] , page 122, lemma 6.12, and use 17.1) the sequence

$$0 \longrightarrow N^D \longrightarrow Y^t \xrightarrow{\phi^t} X^t \longrightarrow 0$$

is exact. The functorial character of this construction follows from the naturality of the isomorphisms used in the generalized WEIL--BARSOTTI formula, and in the CARTIER-SHATZ formula.

QED

COROLLARY (19.2): Let n be a positive integer. There exists a (functorial) isomorphism

$$({}_nX)^D \cong {}_n(X^t)$$

(we write ${}_nX = \mathrm{Ker}(n\cdot 1_X : X \longrightarrow X)$).

This follows, as $(n\cdot 1_X)^t = n\cdot 1_{X^t}$ (cf. 18.3).

As MANIN remarked, his symmetry condition concerning the formal group of an abelian variety follows (over an arbitrary field) from the duality theorem (cf. [23] , page 73, formula 4.8, and page

74, remark 1; symmetry condition: $G_{n,m}$ and $G_{m,n}$ are contained in $\hat{X}$ with the same multiplicity, in case $0 < n < \infty$, $0 < m < \infty$, $(n,m) = 1$; over an algebraically closed field: cf. BARSOTTI, [3], page 84).

COROLLARY (19.3): Let k be a perfect field, $S = \mathrm{Spec}(k)$, and let $\phi : X \longrightarrow Y$ be an isogeny of abelian varieties over k. The dual morphism ϕ^t is separable if and only if

$$(\mathrm{Ker}\ \phi)_{rl} \oplus (\mathrm{Ker}\ \phi)_{ll} = 0.$$

PROOF: Indeed, $\mathrm{Ker}(\phi^t) = (\mathrm{Ker}\ \phi)^D$, hence $(\mathrm{Ker}\ \phi^t)_{loc} = 0$, which is the same as " ϕ^t is separable", and the corollary is proved.

Hence we can give easily an example of a separable isogeny with a purely inseparable dual map. Let X be an abelian variety which has a point of order p (where $p \neq 0$ is the characteristic of the ground field). The canonical homomorphism

$$\phi : X \longrightarrow X/\nu_p$$

is separable, and $\mathrm{Ker}(\phi^t) = \mu_p$ (cf. 2.15), hence ϕ^t is purely inseparable (this remark was already known by CARTIER, BARSOTTI, and many others, I suppose).

COROLLARY (19.4): Let $S = \mathrm{Spec}(k)$, and let X be an abelian variety over k. There exist (functorial) isomorphisms

$$\pi_1(X))^D \cong \pi^1(X^t)$$

$$\pi^1(X))^D \cong \pi_1(X^t)$$

[we write $\pi_1(X)$ for the projective limit of kernels of isogenies over X, and we write $\pi^1(X)$ for the inductive limit of all finite subgroup schemes of X (hence $(\pi^1(X))_{loc} = \hat{X}$); the duality

D: $\underline{N}_k^o \longrightarrow \underline{N}_k$ can be extended:

$$D: (\text{Pro } \underline{N}_k)^o \longrightarrow \text{Ind}(\underline{N}_k)$$

$$D: (\text{Ind } \underline{N}_k)^o \longrightarrow \text{Pro}(\underline{N}_k)] .$$

PROOF: Clearly we can write

$$\pi^1(X) = \varinjlim_n ({}_nX).$$

If $f: X_f \longrightarrow X$ is an isogeny, there exists a commutative diagram

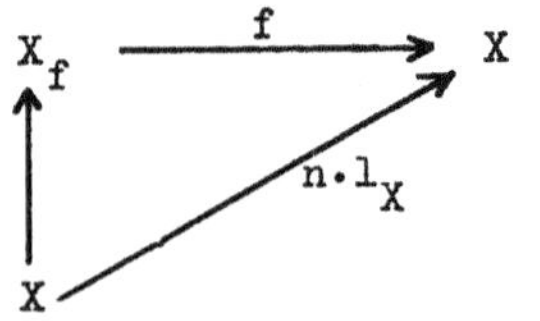

(cf. [41] , VII.52, theorem 27). Hence

$$\pi_1(X) = \varprojlim_n ({}_nX).$$

Thus the corollary follows from (19.2).

COROLLARY (19.5): Let k be a perfect field, and let X be an abelian variety over k. Then

$$\hat{X}_{11}^D \overset{\text{def}}{=\!=} (\varinjlim ({}_nX)_{11})^D \cong \varprojlim ({}_nX^t)_{11} .$$

One could hope to frame the duality theories t and D in a more general concept. However they do not come from one duality theory on a category containing abelian varieties and finite group schemes:

COROLLARY (19.6): Let k be a perfect field, and let

$$0 \longrightarrow Z \longrightarrow X \longrightarrow Y \longrightarrow 0$$

be an exact sequence of commutative k-algebraic group schemes such that X and Y are abelian varieties. Then there exists an exact sequence

$$0 \longleftarrow (CR(Z))^t \longleftarrow X^t \longleftarrow Y^t \longleftarrow (Z/CR(Z))^D \longleftarrow 0$$

("CR(Z) and Z/CR(Z) can be dualized both, but Z itself cannot be dualized in general").

PROOF: Let $T = CR(Z)$, $N = Z/T$, and $U = X/T$. Then $X^t/U^t \cong T^t$ (cf. [22], VIII.4, theorem 10). Fit together the exact sequences

$$0 \longrightarrow N^D \longrightarrow Y^t \longrightarrow U^t \longrightarrow 0$$

and

$$0 \longrightarrow U^t \longrightarrow X^t \longrightarrow T^t \longrightarrow 0.$$

QED

Also the duality cannot be extended to linear algebraic groups: let k be a field of characteristic $p \neq 0$, and $G = \underline{\underline{G}}_{ak}$; there does not exist an algebraic group scheme G^t in this case such that the relations described in (18.6) hold. This is seen as follows: $L_{1,i} \subset \underline{\underline{G}}_a$, hence $L_{1,i}^D = L_{i,1}$ should be contained in G^t, hence the dimension of the tangent space to G^t at the origin should be larger than every natural number, a contradiction.

However it does not seem a good question whether there exists one duality which embodies t and D. Considering the exact sequences (1) and (2) in (19.1), one is inclined to regard t and D as respectively zero- and one-dimensional cohomology operations. This problem perhaps could be settled if we could define and construct higher Picard schemes; we hope to come back to this question in the future.

III.20 The classical duality theorem

In 1959 CARTIER and NISHI proved that for an abelian variety the duality hypothesis holds, i.e. X and X^{tt} are (functorially) isomorphic (cf. [9], [10] footnote; cf. [31]; in [22] this result was used as a hypothesis, see page 216). In this section we show that this result (over arbitrary base preschemes) follows directly from the duality theorem we obtained in section 19.

In this section S denotes a locally noetherian prescheme, and X is an abelian scheme. We assume either that S is an artinian scheme, or that X is projective over S.

Let X^t be the dual of X, and let $\xi_X \in \underline{P}_{X/S}(X^t)$ be the element corresponding to

$$1_{X^t} \in \mathrm{Mor}_S(X^t, X^t) \subset \mathrm{Mor}_S(X^t, \underline{\mathrm{Pic}}(X/S)) \cong \underline{P}_{X/S}(X^t).$$

In section 18 we defined:

$$\sigma = \sigma_{X,X^t/S} : \underline{P}_{X/S}(X^t) \longrightarrow \mathrm{Pic}(X \times X^t);$$

we write

$$\sigma(\xi_X) = \eta_X \in \mathrm{Pic}(X \quad X^t),$$

and we denote by $\eta'_X \in \mathrm{Pic}(X^t \times X)$ the element obtained from η_X after twisting the factors X and X^t. The element

$$\eta'_X \bmod \mathrm{Pic}(X^t) \overset{\mathrm{def}}{=} \rho_X \in \underline{P}_{X^t/S}(X)$$

defines a morphism

$$\kappa_X : X \longrightarrow \underline{\mathrm{Pic}}(X^t/S),$$

as $\underline{P}_{X^t/S}(-) = \mathrm{Mor}_S(-, \underline{\mathrm{Pic}}(X^t/S))$ (the reader will have observed that we take over the usual definitions, cf. [22], page 127).

LEMMA (20.1) (cf. [22]): a) κ_X is functorial in X, it commutes with the zero section, it can be factored

$$X \xrightarrow{\kappa_X} X^{tt} \hookrightarrow \underline{\mathrm{Pic}}(X^t/S),$$

and it is a homomorphism;

b) the following diagram commutes:

$$\begin{array}{ccc} X^t & \xrightarrow{\kappa_{X^t}} & X^{ttt} \\ & \searrow^{1_{X^t}} & \downarrow (\kappa_X)^t \\ & & X^t \end{array} ;$$

c) κ_{X^t} and $(\kappa_X)^t$ are isomorphisms.

PROOF: a) The functorial behaviour of κ is clear from its construction. The element η_X has the property that its restriction to the summand $X^t \subset X \times X^t$ is zero, hence κ_X commutes with zero sections. This implies that it factors through X^{tt} (the fibres of X are connected), and thus κ_X is a homomorphism (cf. [27], page 117, corollary 6.4).

b) From the commutative diagram

$$\begin{array}{ccc} \mathrm{Pic}(X) = \mathrm{Pic}(S \times X) & \xrightarrow{\sim} & \underline{P}_{X/S}(S) \\ \uparrow (0,1)^* & & \uparrow (0_{X^t})^* \\ \mathrm{Pic}(X^t \times X) & \longrightarrow & \underline{P}_{X/S}(X^t), \end{array}$$

we conclude that the restriction of η_X to the summand $X \subset X \times X^t$ is zero. Hence

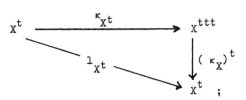

$$\begin{array}{ccccc} \underline{P}_{X^t/S}(X) & \xrightarrow{\ \sigma\ } & \mathrm{Pic}(X^t \times X) & \longrightarrow & \underline{P}_{X/S}(X^t) \\ \kappa_X & & \longmapsto & & 1_{X^t} \end{array} .$$

Thus (b) follows from the diagram

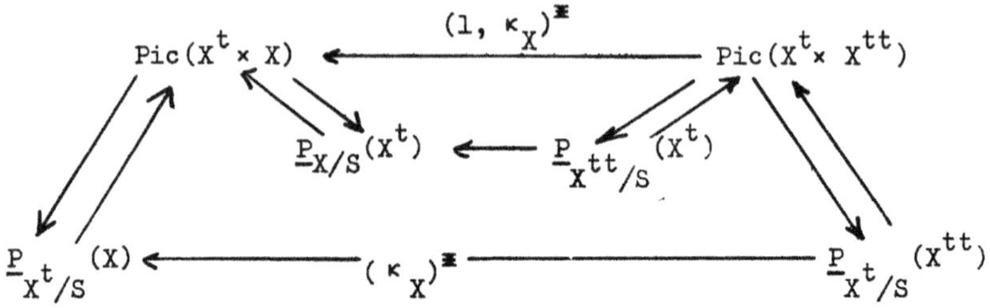

(cf. [22] , page 129).

c) From (b) it follows that κ_{X^t} is monomorphic. Hence it is monomorphic on fibres. For abelian varieties it holds $\dim(Y) = \dim(Y^t)$. Hence κ_{X^t} is surjective on fibres, thus it is faithfully flat (cf. [27] , lemma 6.12). Thus κ_{X^t} is an isomorphism in $\underline{F}_S$ (cf. 17.1b), hence it is an isomorphism.

QED

<u>THEOREM</u> (20.2):

$$\kappa_X : X \xrightarrow{\ \sim\ } X^{tt}$$

is an isomorphism.

<u>PROOF</u>: First we prove this in case $S = \mathrm{Spec}(k)$, where k is a field (CARTIER, NISHI). It suffices to prove this in case k is perfect. Consider the exact sequences

$$0 \longrightarrow K = \mathrm{Ker}(\kappa_X) \longrightarrow X \longrightarrow Y \longrightarrow 0$$

$$0 \longrightarrow Y \longrightarrow X^{tt} \longrightarrow Z \longrightarrow 0$$

(with $(X \twoheadrightarrow Y \hookrightarrow X^{tt}) = \kappa_X$). Then we obtain exact sequences

$$0 \leftarrow (CRK)^t \leftarrow X^t \leftarrow Y^t \leftarrow (K/CRK)^D \leftarrow 0,$$

$$0 \leftarrow Y^t \leftarrow X^{ttt} \leftarrow Z^t \leftarrow 0$$

(cf. 19.6). As $(\kappa_X)^t$ is an isomorphism (cf. 20.1c), it follows that $Z^t = 0$, hence $Z = 0$. Then we obtain an exact sequence

$$0 \longrightarrow K^D \longrightarrow X^{ttt} \xrightarrow{\sim} X^t \longrightarrow 0.$$

Thus $K^D = 0$, and hence $K = 0$.

Now consider the case of an arbitrary (locally noetherian) base prescheme S. From the first half of this proof it follows that $K_s = 0$ for all $s \in S$, and the result follows easily (using Nakayama's lemma). Alternative argument: $\kappa_X : X \longrightarrow X^{tt}$ is surjective on all fibres, by the first half of this proof, thus κ_X is faithfully flat (cf. [27], lemma 6.12), $\mathrm{Ker}(\kappa_X) = K$ is finite and flat over S (cf. section 5), and the sequence

$$0 \longrightarrow K \longrightarrow X \xrightarrow{\kappa_X} X^{tt} \longrightarrow 0$$

is exact (cf. 17.1). Thus by (19.1) the sequence

$$0 \longrightarrow K^D \longrightarrow X^{ttt} \longrightarrow X^t \longrightarrow 0$$

is exact. Hence $K^D = 0$ by (20.1c), thus $K \cong K^{DD} = 0$ (cf. 2.6), κ_X is an isomorphism in $\underline{F}_S$, thus it is an isomorphism.

QED

<u>QUESTION</u>: Consider the exact sequence (1) on page 19 - 1. A commutative diagram with exact rows results:

$$\begin{array}{ccccccccc}
0 & \longrightarrow & N & \longrightarrow & X & \xrightarrow{\phi} & Y & \longrightarrow & 0 \\
 & & \downarrow & & \downarrow{\scriptstyle \kappa_X} & & \downarrow{\scriptstyle \kappa_Y} & & \\
0 & \longrightarrow & N^{DD} & \longrightarrow & X^{tt} & \xrightarrow{\phi^{tt}} & Y^{tt} & \longrightarrow & 0 \; ;
\end{array}$$

thus an isomorphism $N \xrightarrow{\sim} N^{DD}$ results. Is there a connection with the isomorphism mentioned in (2.3)?

REMARK (Gabriel): The situation described in section 19 resembles much (the dual of) the following situation. Consider an exact sequence

$$0 \longrightarrow F \longrightarrow L \longrightarrow N \longrightarrow 0,$$

where F and L are finitely generated, free, commutative groups, and N is a finite group. We write

$$F^D = \mathrm{Hom}_{\mathbb{Z}}(F, \mathbb{Z});$$

clearly

$$\check{N} \overset{\mathrm{def}}{=\!=} \mathrm{Hom}_{\mathbb{Z}}(N, \mathbb{Q}/\mathbb{Z}) \cong \mathrm{Ext}_{\mathbb{Z}}(N, \mathbb{Z}).$$

Thus an exact sequence

$$0 \longrightarrow L^D \longrightarrow F^D \longrightarrow \check{N} \longrightarrow 0$$

results.

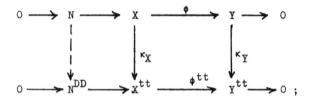

References

1 M.Artin - Grothendieck topologies. Seminar notes, Harvard University, 1962.

2 I.Barsotti - Abelian varieties over fields of positive characteristic. Rend. Circ. Palermo, Ser. II, 5 (1956), 1-25.

3 I.Barsotti - Analytical methods for abelian varieties in positive characteristic. Coll. CBRM, Brussels 1962, 77-85.

4 M.Auslander and O.Goldman - Maximal orders. Trans. Amer. Math. Soc., 97 (1960), 1-24.

5 N.Bourbaki - Algèbre (livre II), chap. 8: Modules et anneaux semi-simples. Hermann, Paris 1958. Act. Sc. Ind. 1261.

6 N.Bourbaki - Algèbre commutative, chap. 1 & 2. Hermann, Paris 1961. Act.Sc. Ind. 1290.

7 D.A.Buchsbaum - A note on homology in categories. Ann. Math. 69 (1959), 66-74.

8 H.Cartan and S.Eilenberg - Homological algebra. Princeton Univ. Press, 1956.

9 P.Cartier - Dualité des variétés abéliennes. Sém. Bourbaki, 10 (1957/58), exp. 164.

10 P.Cartier - Isogenies and duality of abelian varieties. Ann. Math., 71 (1960), 315-351.

11 P.Cartier - Groupes algébriques et groupes formels. Coll. CBRM, Brussels 1962, 87-111.

SGAD M.Demazure and A.Grothendieck - Schémas en groupes. Sém. de Géom. Algébr., IHES 1963.

12 A.Dold - Half exact functors. Lecture notes, Amsterdam 1964 (to appear in: Lect. N. Math.).

13 P.Freyd - Functor categories and their applications to relative homology (multigr. notes, University of Penn., 1962).

14 C.Chevalley - Sur la théorie de la variété de Picard. Amer. J. Math., 82 (1960), 435-490.

15 P.Gabriel - Objets injectifs dans les catégories abéliennes. Sém. Dubreil, Dubreil-Jacotin, Pisot (1958/59), exp. 17.

16 P.Gabriel - Sur les catégories abéliennes localement noethériennes et leurs applications aux algèbres étudiées par Dieudonné. Sém. J.-P.Serre, 1960.

17 P.Gabriel - Des catégories abéliennes (thèse Sc. math. Paris, 1961). Bull. Soc. Math. France, 90 (1962), 323-448.

SGAD P.Gabriel - Construction des préschémas quotient/ Généralités sur les groupes algébriques. Sém. de Géom. Algébr., IHES 1963, exp. 6/7.

18 A.Grothendieck - Sur quelques points d'algèbre homologique. Tôhoku Math. J., 9 (1957), 119-221.

SGA A.Grothendieck - Séminaire de Géom. Algébr., IHES, 1960/61.

EGA A.Grothendieck and J.Dieudonné - Éléments de géométrie algébrique. I,II,III^1,III^2,IV^1,IV^2: Publ. Math. No. 4 (1960), 8 (1961), 11 (1961), 17 (1963), 20 (1964), 24 (1965).

FGA A.Grothendieck - Fondements de la géométrie algébrique (extraits du Sém. Bourbaki 1957-1962). Paris, 1962.

19 R.Hartshorne - Connectedness of the Hilbert scheme (thesis, Princeton University, 1963).

20 J.-I.Igusa - Arithmetic variety of moduli for genus two. Ann. Math. 72 (1960), 612-649.

21 E.R.Kolchin - On certain concepts in the theory of algebraic matric groups. Ann. Math. 49 (1948), 774-789.

22 S.Lang - Abelian varieties. Intersc. Publ., New York, 1959.

23 Yu.I.Manin - The theory of commutative formal groups over fields of finite characteristic. Russian Math. Surveys, 18 (1963), 1-80 (= Uspehi mat. Nauk 18 (1963), 3-90).

24 H.Matsumura and M.Miyanishi - On some dualities concerning abelian varieties (to appear in Nagoya Math. J.).

25 M.Miyanishi - On the extensions of linear groups by abelian varieties over a field of positive characteristic p (to appear in J. Math. Soc. Japan).

26 D.Mumford - Lectures on curves on an algebraic surface (with the assistence of G.M.Bergman). Lecture notes, Harvard University, 1964.

27 D.Mumford - Geometric invariant theory. Ergebnisse Math., Bd. 34. Springer Verlag, 1965.

28 J.P.Murre - On contravariant functors from the category of preschemes over a field into the category of abelian groups (with an application to the Picard functor). Publ. Math. No 23, IHES, 1964.

29 J.P.Murre - Representation of unramified functors. Applications. Sém. Bourbaki 17 (1964/65), exp. 294.

30 M.Nagata - Local rings. Intersc. Publ., New York, 1962.

31 M.Nishi - The Frobenius theorem and the duality theorem on an abelian variety. Mem. Coll. Sc. Kyoto, 32 (1959), 333-350.

32 D.G.Northcott - Ideal theory. Cambridge Univ. Press, 1960.

33 F.Oort - Yoneda extensions in abelian categories. Math. Ann., 153 (1964), 227-235.

34 M.Rosenlicht - Extensions of vector groups by abelian varieties. Amer. J. Math., 80 (1958), 685-714.

35 J.-P.Serre - Espaces fibrés algébriques. Sém. C.Chevalley, 2 (1958), exp. 1.

36 J.-P.Serre - Quelques propriétés des variétés abéliennes en caractéristique p. Amer. J. Math., 80 (1958), 715-739.

37 J.-P.Serre - Sur la topologie des variétés algébriques en caractéristique p. Symp. Intern. Top. Alg. Mexico, 1958, 24-53.

GA J.-P.Serre - Groupes algébriques et corps de classes. Hermann, Paris, 1959. Act. Sc. Ind. 1264.

GP J.-P.Serre - Groupes proalgébriques. Publ. Math. No 7, IHES, 1960.

38 S.S.Shatz - Cohomology of artinian group schemes over local fields. Ann. Math. 79 (1964), 411-449.

39 A.-M.Simon - Schémas de Grothendieck et de Chevalley. Multigr. notes, Univ. libre de Bruxelles, 1963/64.

SGAA J.-L.Verdier - Topologies et faisceaux. Sém. de Géom. Algébr., IHES, 1963/64, exp. 1-4.

40 B.L.van der Waerden - Algebra II. Springer Verlag, 4th Ed., 1959.

41 A.Weil - Variétés abéliennes et courbes algébriques. Hermann, Paris, 1948. Act. Sc. Ind. 1064.

42 N.Yoneda - On the homology theory of modules. J. Fac. Sc. Tokyo, 7 (1954), 193-227.

43 N.Yoneda - On Ext and exact sequences. J. Fac. Sc. Tokyo, 8 (1960), 507-576.

44 S.Mac Lane - Homology. Springer Verlag, 1963.

.ecture Notes in Mathematics

ts erschienen

ıar über Funktionen-Algebren
ɔhn Wermer, Professor an der Brown University, Providence
SA. (Gehalten an der Eidg. Technischen Hochschule,
hungsinstitut für Mathematik, Zürich, Wintersemester
64)
tscher Sprache. IV, 30 Seiten. 1964. Steif geheftet DM 3,80

mologie des espaces localement compacts d'après J. Leray
rmand Borel, Institute for Advanced Study, Princeton N.J.
sés faits au Séminaire de Topologie algébrique de l'Ecole
chnique Fédérale au printemps 1951)
age. In französischer Sprache. IV, 93 Seiten. 1964.
jeheftet DM 9,–

e Homotopy Theory
rank Adams, Department of Mathematics, University of
ester. (Lectures delivered at the University of California at
ley 1961. Notes by A.T. Vasquez)
ischer Sprache. IV, 74 Seiten. 1964.
eheftet DM 7,80

s of Homotopy Classes
ormulas and homotopy-commutativity
Arkowitz and C.R. Curjel, Dartmouth College, Hanover, N.H.,
sity of Washington, Seattle, Wash., Forschungsinstitut für
matik, Eidg. Techn. Hochschule, Zürich
ischer Sprache. IV, 36 Seiten. 1964. Steif geheftet DM 4,80

nologie Galoisienne
an-Pierre Serre, Collège de France, Paris (Cours au Collège
nce, 1962 – 1963)
age. In französischer Sprache. VIII, 214 Seiten. 1965.
eheftet DM 18,–

ermlogik mit Auswahloperator
ans Hermes, Universität Münster, Institut für mathematische
nd Grundlagenforschung.
scher Sprache. IV, 42 Seiten 1965.
eheftet DM 5,80

ction to Lie Groups and Transformation Groups
ippe Tondeur, Department of Mathematics, University
ch
scher Sprache. VIII, 176 Seiten. 1965. Steif geheftet DM 13,50

elliptic differential systems and eigenvalue problems
ns Hopkins University, Baltimore Md, March-May 1965
tano Fichera, University of Rome
scher Sprache. IV, 176 Seiten. 1965. Steif geheftet DM 13,50

Heft 9
Pseudo-Boolean Programming and Applications
Presented at the Colloquium on Mathematics and Cybernetics in the Economy, Berlin, October 1964
By Petru L. Ivănescu, Institute of Mathematics, Romanian Academy of Sciences, Bucharest
In englischer Sprache. IV, 50 Seiten. 1965. Steif geheftet DM 4,80

Heft 10
Die Suzukigruppen und ihre Geometrien
Vorlesung Sommersemester 1965 in Mainz
Von Heinz Lüneburg, Mathematisches Institut der Universität Mainz
In deutscher Sprache. VI, 111 Seiten. 1965. Steif geheftet DM 8,–

Heft 11
Algèbre Locale · Multiplicités
Cours au Collège de France, 1957–1958, rédigé par Pierre Gabriel
Par Jean-Pierre Serre, Collège de France, Paris.
2. Auflage. In französischer Sprache. VIII, 192 Seiten. 1965.
Steif geheftet DM 12,–

Heft 12
Halbexakte Homotopiefunktoren
Von Albrecht Dold, Mathematisches Institut der Universität Heidelberg
In deutscher Sprache. II, 157 Seiten. 1966. Steif geheftet DM 12,–

Heft 13
Seminar on Fiber Spaces
Lectures delivered in 1964 in Berkeley and 1965 in Zürich. Berkeley notes by J.F. McClendon.
By Emery Thomas, University of California, Berkeley, Forschungsinstitut für Mathematik, ETH, Zürich
In englischer Sprache. IV, 45 Seiten. 1966.
Steif geheftet DM 4,80

Heft 14
Vorlesung über Approximationstheorie
Westfälische Wilhelms-Universität, Münster
Institut für Numerische und Instrumentelle Mathematik, Sommer-Semester 1964
Von Helmut Werner. In deutscher Sprache. IV, 196 Seiten. 1966.
Steif geheftet DM 14,–

Heft 16
Compact Systems of Sets
By Johann Pfanzagl and Winfried Pierlo, Seminar für Wirtschafts- und Sozialstatistik an der Universität zu Köln
In englischer Sprache. Etwa IV, 48 Seiten. 1966.
Steif geheftet DM 5,80

In Vorbereitung

Heft 17
Spherical Harmonics
Von Claus Müller, Institut für Reine und Angewandte Mathematik Technische Hochschule Aachen
In deutscher Sprache. IV, 45 Seiten. 1966. Steif geheftet etwa DM 5,–